Social Organization of Hamadryas Baboons

SOCIAL ORGANIZATION OF HAMADRYAS BABOONS

A FIELD STUDY

HANS KUMMER

The University of Chicago Press
Chicago and London

Published simultaneously as
BIBLIOTHECA PRIMATOLOGICA, No. 6
Editors: H. Hofer, Covington, La.; A. H. Schultz, Zürich; D. Starck, Frankfurt a. M.

The University of Chicago Press, Chicago and London, S. Karger AG, Basel
The University of Toronto Press, Toronto 5

Library of Congress Catalog Card Number: 67-25082. Printed in Switzerland

CONTENTS

ACKNOWLEDGEMENTS

This study was made possible by a post-doctoral fellowship to the author from the Swiss National Foundation for Scientific Research (Schweizerischer Nationalfonds für wissenschaftliche Forschung) as well as by substantial contributions from the George and Antoine Claraz Fund and the Swiss Society for Nature Study (Schweizerische Naturforschende Gesellschaft). I am grateful also to the Wenner-Gren Foundation for a grant in aid of preparing the manuscript. I wish to thank Professors H. HEDIGER, E. HADORN and A. SCHULTZ of the University of Zurich, and Professors D. STARCK and H. FRICK of the University of Frankfurt a. Main, who have generously given us their aid and advice, and Drs. CLAIRE and WILLIAM RUSSELL, WILLIAM MASON, E. W. MENZEL and S. ALTMANN for reading the manuscript and for their valuable comments.

We owe special thanks, furthermore, to Mr. J. DE STOUTZ and Mr. R. HEINIS of the Swiss Legation in Addis Abeba for their efforts on our behalf. We are indebted to the ministries of the Imperial Ethiopian Government for their good will towards our project and to the many residents of Ethiopia who helped us with information concerning the whereabouts of the hamadryas. Not the least of our thanks and gratitude goes to our 'bodyguard' of the Ethiopian police, and to the Gurgure people for their hospitality.

Needless to say, the greatest part of my thanks is due to my assistant and colleague, FRED KURT, for his tireless efforts during the work in the field.

LEVELS OF SOCIAL ORGANIZATION IN HAMADRYAS BABOONS

Troop (a–c)
A troop consisting of three bands, females and young omitted. The bands become apparent in three contexts:

(a) Separate choice of direction before departure from the sleeping-rock.
(b) Separate columns after the common departure.
(c) Fights between the bands.

Band (d)
Females and young omitted.

Two-male team (e–j)
Developmental stages from left to right, in typical travel formation. Younger male and his females in dark color. Offspring omitted.

(e) One-male unit with subadult follower.
(f) One-male unit with adult follower.
(g) The follower has formed his own initial unit.
(h) Mature stage: Younger male initiates, older male directs travel.
(j) Late stage: Older male has lost his females, but still directs travel.

One-male unit (k–n)
Developmental stages from left to right. Offspring omitted.

(k) Initial unit.
(l) Transitional stage.
(m) Mature stage.
(n) Late stage.

Drawings by Carl Cramer

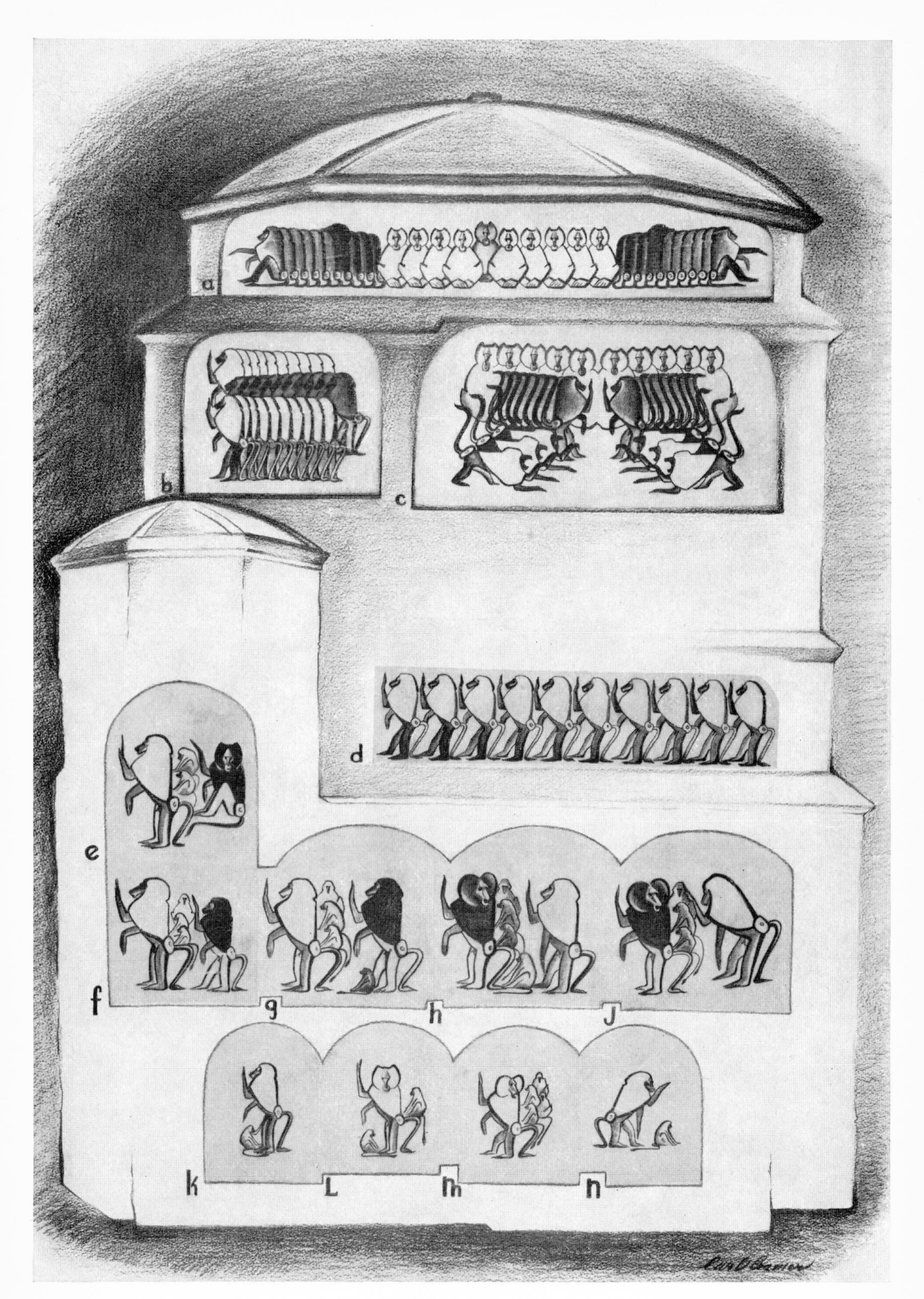
a
b
c
d
e
f
g
h
J
k
L
m
n

'They are considered the lowest of the Catarrhine monkeys, and as they are generally of large size they are dangerous animals when adult, possessing savage and ugly dispositions.'

Elliot, p. 116 (1913)

I. INTRODUCTION

Baboons have adapted to a variety of habitats ranging from West African rain forests to semidesert areas on the coast of the Red Sea. While all baboons are morphologically adapted to life on the ground, some species have become more independent of trees than others. In a rough ecological series, we find on one end the forest dwelling West African species *(Mandrillus leucophaeus, M. sphinx, Papio papio)*, none of which has so far been studied in the field. The first step into open country is realized by the savanna baboons of South and East Africa, including, from south to north, the species *Papio ursinus, cynocephalus* and *anubis*[1]. Their social organization and its ecological context have been subject to long range field work by HALL (1962 a, b) in South Africa, by DEVORE (1962) in Kenya, and by ALTMANN and ALTMANN (in preparation) in Kenya and Tanzania. In both regions, the groups can range far into the open grassland, but at night, they withdraw to high trees or, as in the Cape region, to vertical cliffs.

In the dry north-eastern extreme of the generic area, the savanna baboons are eventually replaced by the desert baboon, *Papio hamadryas*, which lives in the arid brush land of eastern Sudan, the eastern lowlands of Ethiopia and the Somali lands. The south-western boundary of its distribution runs approximately from Port Sudan to the lower course of the Webi Shebeli River (STARCK and FRICK, 1958). Separated from this main area by the Red Sea, hamadryas baboons are also found in South West Arabia. In typical hamadryas habitats, tall trees are scarce, and the baboons pass the night on vertical rock-faces. Judging from the comparatively low population density and the long daily travels of hamadryas baboons, the food supply in their habitats is even more meager than in the regions where the savanna baboons have been studied.

The baboons exemplify that even closely related species, which use a very similar set of communicative gestures and vocalizations,

[1] The taxonomic terms are used in concordance with the recommendations of the Committee on Baboon Taxonomy (OSMAN HILL, in preparation).

may nevertheless form essentially different societies. In 1932, ZUCKERMAN reported that the hamadryas baboons of the London Zoo lived in a system of family units, in which each male owned and defended a harem of females. ZUCKERMAN assumed that the harem organization was typical of the 'subhuman level' of primates in general. When, thirty years later, WASHBURN and DEVORE (1961 a, b) studied savanna baboons in Kenya, they were again tempted to generalize their results on taxonomic grounds. They concluded that 'there is nothing resembling a family or a harem in baboons.' As it turned out, hamadryas baboons live in family units, whereas savanna baboons do not.

The common social organization among old world monkeys, including the savanna baboons, consists of stable groups with many adult males. Within a group, all males potentially have access to all females, although the more dominant males usually perform most of the reproductive activity. Three species of three different genera have so far been found to deviate from this general pattern in that they have adopted, as the standard social unit, the one-male group consisting of several females constantly and exclusively associated with one male. This pattern is found in the patas monkey, *Erythrocebus (Cercopithecus) patas*, of Uganda (HALL, 1965); CROOK has found it in the gelada *(Theropithecus gelada)* of the Ethiopian mountains (CROOK and GARTLAN, 1966); and in the genus *Papio*, the one-male unit pattern appears in *P. hamadryas* (KUMMER and KURT, 1963). All three species live in open habitats, where trees are rare and too small to protect the monkeys from predators. Since the three specialists of the open country have evolved independently in different genera, their one-male units may be convergent adaptations to similar environmental factors which, however, are not yet identified.

Patas monkeys, geladas and hamadryas baboons are further characterized by a pronounced sexual dimorphism. The adult males weigh about twice as much as the females; the gelada and hamadryas males, in addition, have mantles of long hair; and in the hamadryas the males also differ from the females in color. The same order of species is apparent in the intensity with which the male leads and herds his females. Such control is virtually absent in the patas and most pronounced in the hamadryas. The exact function of the male's appearance in the formation of one-male units remains to be investigated. There are indications of a correlation between the occurrence of polygyny and excessive sexual dimorphism in both size and other aspects of external morphology among pinnipeds and some other vertebrates. There is also a suggestion of a relationship with habitat, such that polygyny is more common in animals occupying open areas. Quantitative verification is still necessary on both these points (MARLER, personal communication).

The savanna dwelling patas monkeys avoid predators on the ground by rapid flight and by elaborate hiding techniques. Probably in relation to this, their one-male units live far apart from each other. In contrast, geladas and hamadryas, at least in their contacts with dogs, follow the savanna baboon pattern of coordinated defense on the ground, and their one-male units live together in large troops. In the gelada and the hamadryas we thus find an organization which is unique among primates. Small, stable units of a family type are integrated into larger troops. Such a two-level system obviously faces the problem of maintaining the integrity of the smaller units within the larger.

Prior to the present study, hamadryas behavior had not been investigated in the wild. Captivity studies had, however, been made by ZUCKERMAN (1932) on the colony of the London Zoo and by the author (1957) on a smaller group at the Zurich Zoo. In both colonies, the baboons established a system of one-male units comparable to the organization in the wild. The differences in the behavior of the wild and the captive groups have been described in an earlier paper (KUMMER and KURT, 1965). In the Russian primate research station of Sukhumi, hamadryas baboons have been bred for several generations. Unfortunately, no information on the persistence of their social organization was available to me. The only other scientific reports on hamadryas cover matters other than group behavior. HEINROTH (1959) raised hamadryas infants in her home. A charming report by MACDONALD (1959) on home-raised Arabian hamadryas includes an account and bibliography of the domesticated hamadryas of ancient Egypt where, at one period, the 'Sacred Baboon' was venerated as the incarnate of Thot, the God of Scribes and Scholars.

The present paper reports on a field study that was carried out in Ethiopia from November, 1960 to October, 1961 by FRED KURT, a student of Zoology at the University of Zurich, and myself.

1. AIMS AND METHODS

Our field study was a preliminary one. Its immediate aim was to provide a view on the entire system to be studied, i.e., on the hamadryas population in the type of environment that must have affected the latest steps of its evolution. Our purpose was not to test but to formulate hypotheses, especially on the function of behaviors and grouping tendencies in the entire organization[1]. This search for relevant topics led us to try a variety of research procedures, and explains the lack of thoroughness in the treatment of specific topics.

The most obvious characteristic of a 'social' species is that its members, instead of being evenly distributed in their habitat, form mobile concentrations. Such super-individual systems or social units were the object of this study. They were described first by the spatial arrangements of the individuals, and secondly, by the gestures and vocalizations exchanged. These two classes of phenomena have also been used as measures for our working definition of a social unit (Kummer and Kurt, 1963). A list of the most important social gestures and sounds is given on page 180. A more comprehensive description was presented in my zoo study (1957).

From a number of quantitative recording programs tried out in the first months of the present study, we only carried out the most effective. These were found to answer, at least partially, three sets of questions. The first is concerned with the arrangement of the population in space:

1. Can certain units within the hamadryas population be distinguished on the basis of their spatial detachment?

2. What is the composition of these units, in terms of number, sex, and ages of the individuals?

3. How are members of a unit ordered spatially within the unit?

This anatomical aspect must be understood in the context of a functional order, which can be approached by a second set of questions:

4. How frequent and of what sort are the observable signals, i.e., gestures and vocalizations, exchanged between members of the same and of different detached social units?

[1] Significantly, the organizing function of the neck bite (p. 36) and of notifying behavior (p. 128) were not understood in the captivity study, although they were regularly observed.

5. From information on the effects of these signals, what conclusions may be drawn concerning the social functions or the 'roles' of the members and therewith about the functioning order of the entire social unit?

A third set of questions that can be answered in part by descriptive methods concerns the life cycles of individuals and of social units:

6. How do the social 'roles' change during the life of an individual?

7. Is there a life cycle of the social unit?

8. What can these findings tell us about the maintenance of social organizations over the generations?

The answers to these questions, and all basic questions beyond them, will eventually require the use of experimental methods. One standard type of biological experiment, that of transplantation, is indispensable in studying environmental effects on all levels of an organism. As cells and organs are transplanted from one individual to another, at various stages of development, so individuals varying in age, or entire social units, should be transplanted into units of different social organization or explanted into different environments. Only such methods can reveal a species' genetic potentials for social organization, and identify the modifying influence of the physical and social environment on each generation. The present study includes a short pilot study on the feasibility and the immediate effects of such transplantations in the field.

2. TECHNIQUES

In order to get the animals accustomed to us, we avoided all hiding places and showed ourselves as often and as openly as possible. As soon as the baboons retreated at our approach, we would also retreat a few paces, then sit down and avoid looking directly at them. This technique usually reduced the flight distance from about 200 to 50 meters on the first day. After six months, the flight distance amounted to some 40 to 80 meters in open terrain and between 20 and 30 meters at the sleeping rock. As far as we could tell, the animals were only affected by us in the following ways: In the first weeks, we sometimes burst unintentionally into the middle of a party. During the entire year, two and three year old males would occasionally threaten us by staring at us and slapping the ground. Loud noises, such as the warning call of a bird or the rolling of a stone, sometimes caused a few individuals to turn around and look at us. Our presence protected the animals from predators since these kept further away from us than did the baboons. On three occasions during the last month, we placed food on a sleeping cliff before the arrival of the animals in order to artificially provoke fighting. On four occasions at the end of the field study, we realesed marked animals from a distant troop into the troop under close observation.

In identifying individuals we relied mainly on facial and anal region scars. Also used, though less suitable, were the shapes of the nostrils and supraorbital ridges, as well as the shape of the tufts at the end of the tail. The relative lengths and the position of the nipples were useful in identifying adult females. Infants belonging to identified mothers were mostly characterized only by their size and sex.

The baboons were followed on foot. Observations were made with 6 × 24 field glasses or with the naked eye. Protocols were taken in notebooks. Time was recorded every 1 to 3 minutes during continuous observation. In recording the composition of moving troops, one of us would take the count through field glasses while the other jotted down the figures. Most photographs were taken with Pentax 35 mm cameras and a 500 mm telelens. Unfortunately, light conditions in the sleeping cliffs were often minimal. Sleeping rocks, as well as the animals' travel routes, were mapped trigonometrically with a pair of field compasses. In all, 835 hours were spent in observation; of this 422 were spent at the sleeping rock and 413 in open country. On the average, about 80 animals were visible at the sleeping rocks, while in the field the average was about five. Capturing and marking techniques are described on page 108.

3. SAMPLES

We studied the hamadryas population in three concentric samples of decreasing size (Fig. 1) and with increasing attention to details. Although the data collected from the samples differed in type, they overlapped sufficiently to permit cross-sample verifications on the more important structures, in analogy to the pictures seen through the different lenses of a microscope.

a. *The survey*. This largest sample comprised 23 troops along a line of 340 kilometers in length extending from east to west. This

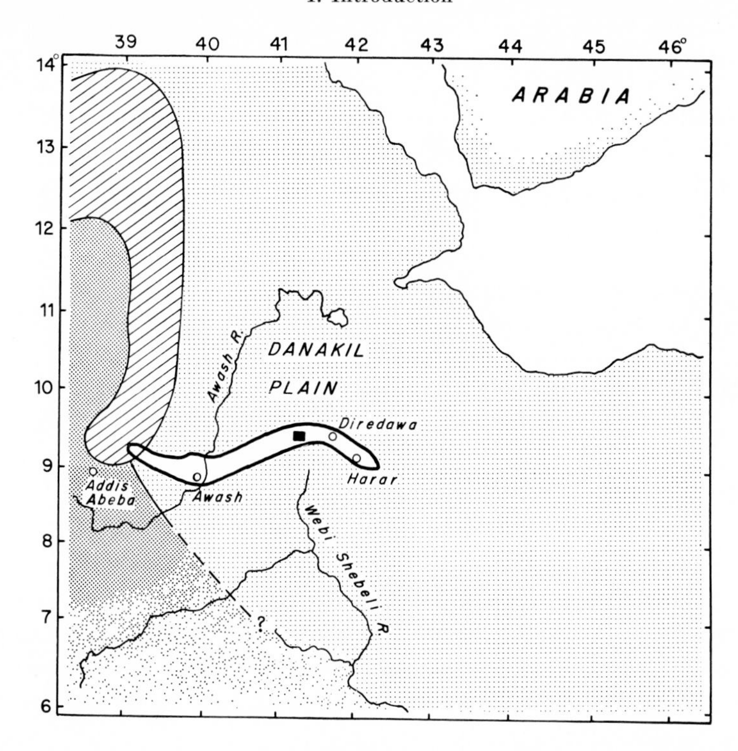

Fig. 1. Geographical distributions of *Papio hamadryas* (light dots), *Papio anubis* (heavy dots) and *Theropithecus gelada* (bars), after STARCK and FRICK (1958). The heavy line marks the area of our survey, the black rectangle the area of the broad sample.

represents approximately one quarter of the entire east-west distribution of the species. The survey served primarily as the basis for studying the variability of the biotope, and of the size and composition of the troops.

b. *The broad sample.* We next chose an area of 20 × 25 kilometers from the survey for closer investigation (Fig. 11). This area, which extends to the north and east of the village of Erer-Gota, contained 10 sleeping rocks and a population of about 700 animals; we also included another troop, residing at Diredawa, outside this area. Size and age-sex composition of the troops was determined more exactly than was possible in the survey. We also studied the spatial segregation of the troop into smaller units and the relative distances between the members of different sex-age classes, and we compared these measures with the number and types of social interactions between the classes. Individuals were not identified. In three troops, we made preliminary studies of the decamping procedures in the morning and the form of the daily routes.

c. *The close study*. From the troops included in the broad sample, the White Rock Troop near Garbelucu appeared to be most representative, and as it was technically acceptable it was chosen for a close study. The processes of decamping and troop movement were analyzed in more detail. Eight one-male units of this troop, as well as two one-male units of another troop, were identified and observed over periods from one to six months. Altogether, we knew 41 individual animals.

4. TERMS FOR SOCIAL UNITS AND SEX-AGE CLASSES

To make a reading of the following chapters easier, I shall begin with a rough outline of the social structure of a hamadryas population and a description of its daily routine. Throughout the survey area, extending from Sendafa near Addis Abeba in the west to Harar in the east, all hamadryas were found to spend their nights sleeping on vertical cliffs: these we shall call *sleeping rocks*. The largest social unit found in hamadryas baboons is the *troop*. It consists of all animals sleeping together on a particular cliff. These troops of one hundred or more animals are not constant in their make-up. Under certain conditions, a troop may split up into several *bands* which consist of 30 to 90 animals. Each band is composed of several *one-male units*, comprised of an adult male and one to nine females with their young. Frequently two adult males cooperate in leading their one-male units. Such pairs of unit leaders we shall call *two-male teams*. All females live in one-male units, whereas young males between two years of age and early maturity are only loosely, if at all, attached to such units. Subadult and young adult males who do not themselves lead units, but are nevertheless always found in the company of a unit belonging to an older male we shall call *followers*. In his Zoo study, ZUCKERMAN (1932) called the one-male units 'harems' and the followers 'bachelors.' As a loose descriptive term for momentary gatherings of animals, the constancy as well as the structure of which are unknown, we will use the word *party*. By *class* we mean individuals of the same age and sex. A summary of the classes we have distinguished is given in Table I.

Table I. Sex-age classes and their characteristics.

Class	Estimated age in years	Characteristics
Adult males	More than 7	Weights of two individuals 18.0 and 18.5 kg. Sitting height about 65 cm. Fully developed silvery, salt and pepper mantle. *Younger males* with blackish hair on arms, deep red anal field, darker face color; hair coloring along middle of the back brown, especially in the sacral region. Long upper canines. *Older males* no longer have brown hair. Face bright red. Snout appears longer, canines are worn.
Subadult males	5–7	Weight of one individual 16 kg. Sitting height about 60 cm. Mantle completely developed. With the exception of the light tufts of hair at the sides of the head still primarily brown.
	4–5	Sitting height 56 cm. Mantle half developed, brown. Snout relatively short.
	3½–4	Sitting height about 50 cm. First signs of mantle appear in the shoulder region.
Three-year-old males	2½–3½	Weight of one 2½ year old 5.5 kg. Sitting height about 47 cm. No signs of mantle, but in some cases longer hairs at the sides of the head. At greater distances, easily confused with adult females.
Two-year-old males	1½–2½	Sitting height about 40 cm. Brown hair.
One-year-old males	½–1½	Sitting height about 30 cm. Brown hair.
Black males	0–½	Hair completely or partially black.
Adult females	More than 5	Weights of 2 individuals 8.7 and 10 kg. Sitting height about 55 cm. Nipples always stretched, longer than wide.

Table I continued

Class	Estimated age in years	Characteristics
Sub-adult females	3 ½–5	Weight of 2 individuals 6.6 and 8 kg. Sitting height about 50 cm. Nipples usually short, not longer than wide.
Three-year-old females	2 ½–3 ½	Sitting height about 45 cm. Nipples always button-like.
Two-year-old females		
One-year-old females		Same as for males of equal age.
Black females		

Ages of young animals were estimated on the basis of my previous experience with hamadryas at the Zurich Zoo where animals of known ages were observed through adulthood. Estimates of age made by Kurt never differed from mine more than ½ year for infants and juveniles and never more than one age class for older animals. At the end of our field work, as a check on the accuracy of our age estimates, we collected 18 animals at Diredawa and estimated their respective ages as before on the basis of the above criteria, without reference to their dentition. This sample also provided the weights and measures given in Table I. Subsequently, the skeletons were examined at the Anthropological Institute of the University of Zurich (Prof. J. Biegert). On the basis of tooth eruption, the ages were, of the infants, juveniles and subadults, determined according to the tables by Hurme and van Wagenen (1962) for the rhesus monkey and by Reed (1965) for savanna baboons. Our estimates deviated from these figures on an average of 0.6 years. In the worst case the difference was 1.5 years.

5. DAILY LIFE IN THE HAMADRYAS TROOP

In hamadryas troops social and nonsocial behavior alternate in a daily routine which remains essentially unchanged throughout the year. Though eating and drinking habits as well as reproduction are

Fig. 2. A troop at dawn. Some animals are still sitting on the sleeping ledges, while others have shifted to the waiting areas above the cliff.

Fig. 3. Dozing and social activity above the cliff in the morning. Hamadryas occasionally rest on trees during the day; at night they strictly prefer cliffs.

subject to seasonal changes, the structure of the troop and the daily march from the sleeping rock into the savanna remains constant.

At about sunrise the parties leave their ledges in the sleeping rocks and go to the open waiting areas above the face of the cliff to sit in the sun (Fig. 2). Usually these areas are comprised of more or less bushless slopes above the sleeping rock. If the area above the rock is not sufficiently large (ca. 30 meters in diameter) and bare, the animals may move out into some open space in the vicinity of the sleeping rock and remain there until their departure. The time spent in waiting areas can vary from a few minutes to as much as three hours. During this time the adult animals will doze, heads sunk on their chests, or will devote themselves to social grooming within the one-male unit (Fig. 3). Chasing between adult males and copulation primarily occur during this morning rest period, while young animals form play groups. At the same time the troop is preparing for its departure: The shifts of the individual one-male units become more and more frequent. Gradually the outlines of the troop change and, as from an amoeba, pseudopods are sent out and drawn in again until finally one of these grows longer and longer and the rest of the troop flows into it.

The pseudopod now becomes a column which leaves the area in a quick march (Fig. 4). Male chasing, grooming, play and copulation stop abruptly and, except for grooming, are not again resumed, even during occasional rests in the savanna. At first, the troop stays together closely, moving rapidly along open ridges or river beds, without stopping to eat or drink. After about half an hour to an hour the speed decreases, the column becomes looser and individual parties turn off from the direction of the march, beginning to eat. Before midday the main body of the troop usually comes to a halt and the animals spread out over an area as much as a kilometer in diameter. Now they eat and rest in individual units (Fig. 5). During the dry season the bands will usually come together at midday at some shady riverbed which still contains water (Fig. 6). Drinking is accompanied by a one to two hour rest period. Grooming and play occur at this time but chasing between males and copulation is as a rule omitted. In the rainy season this midday rest is replaced by a number of short rest periods under bushes while the animals drink here and there as they cross streams and rivulets.

The return march by the various portions of the scattered troop often begins even before the midday rest. Individual parties of the

Fig. 4. The common departure from the sleeping rock is the only activity in which the troop acts as a unit. A dry river bed is often used for this fast and compact travel.

Fig. 5. The troop has split up into small foraging parties. The leader of a resting one-male unit watches as another unit passes by.

Fig. 6. Midday rest during the dry season. A one-male unit is drinking from the river.

troop may lose contact with the main body entirely and will, perhaps, spend the night on another sleeping rock. Most of the troop, however, begin to wander back toward the rock they left that morning, eating leisurely as they move. Towards evening the pace usually quickens again and the troop arrives at the sleeping rock in two or three large bands or smaller parties. The first parties usually arrive between 15.30 h and 16.30 h, and return to the waiting area where the complete range of social activities is resumed, including male chases and sexual behavior. The last parties arrive as much as two hours after the first and, passing through the waiting area, go directly onto the rock. Before dusk all of the animals enter the sleeping rock, the one-male units regularly returning to the same ledges each night. Social behavior now wanes gradually and with the onset of darkness all the animals are dozing or sleeping on their narrow ledges (Fig. 7 and 8), except for a few young animals which continue to chase each other about the cliff.

Fig. 7. A one-male unit sleeping on its ledge in the rock face.

Fig. 8. A party of subadult males asleep. Members of one-male units sleep closer together and rarely lie down.

II. THE SURVEY

The largest or survey sample extends from Sendafa near Addis Abeba (39° 10′ Eastern Longitude) to the Erer Valley east of Harar (42° 13′ Eastern Longitude) (Fig. 1). From west to east, this area covers 348 km in length. Beginning at 9° Northern Latitude, it extends northwards for 75 km. In this strip 23 troops were counted and observed for a few hours each in order to obtain some information on the variations of troop size and composition and on ecological conditions. A further aim was to make observations in the zones of contact between *Papio hamadryas* and their neighbors in the west, *Papio anubis* and *Theropithecus gelada*.

By asking local residents to lead us to sleeping rocks (amharic 'djindjero bed,' i.e., monkey house), we were usually able to find hamadryas troops quickly. Once at the sleeping rock, a count of the incoming troop would be taken in the evening; another the next morning as it left. The troop would then be followed for several hours while we took note of its biotope, its behavior towards observers, and as much as possible of its social structure.

1. VARIATIONS IN ALTITUDE AND VEGETATION

The 23 samples were taken at altitudes ranging from 820 meters above sea level at Alibete, north of Awash station, and 2180 meters at Aliltù near Sendafa. This second area, however, is not constantly occupied by hamadryas.

A rough idea of the varying density of vegetation in different hamadryas areas is given in Table II which is based on estimates made from hill tops in 17 of the samples. Typically, the hamadryas biotope is poor grassland dotted with shrubs and acacia trees 5 to 10 meters apart and 3 to 5 meters in height. In areas where fields are cultivated, such as at Aliltù, the animals feed on grain and field produce; here they are driven off with stones, and they are hunted regularly. In areas without agriculture, as in the neighborhood of Erer-Gota, the diet consists chiefly of grass ears and the flowers, leaves and fruits of shrubs and trees.

Table II. Percents of surface covered by vegetation of different density in 17 hamadryas biotopes.

Type of vegetation	Mean	Highest proportion	Smallest proportion
Bare earth	5.2%	20%	0%
Grassland with neither brush nor trees	1.8%	30%	
Cultivated fields	15.5%	65%	0%
Shrubs, 10 to 50 meters apart	14.3%	90%	
Shrubs, 5 to 20 meters apart	38.4%	90%	35%
Shrubs, less than 3 meters apart	25.2%	80%	

A special relationship was found between the hamadryas and the opuntia cactus growing in the region of Diredawa. The fruit of the cactus are eaten in quantity. In the bush, the opuntia is rare; but, around the villages the cactus form thick prickly belts. A similar belt, which we ventured to pass but once, surrounds a hamadryas sleeping rock miles away from the next village, but not the smaller uninhabited cliffs in its neighborhood. In the niches of this sleeping cliff, the cactus seeds from the baboon feces form thick layers. A portion of another huge troop near Diredawa has adopted the habit of spending its nights on flat ground using the opuntias as its only protection (see right portion of Fig. 2). Thanks to the loose symbiosis between the baboon and the recently introduced opuntia, the hamadryas of this region may eventually become independent of the scarce sleeping rocks by planting such protective screens.

In all the biotopes we found few, if any, trees which were strong and large enough to provide a suitable sleeping place for the animals. All the troops slept in nearly vertical rock-faces, none on trees. Even in flight from the observer, the hamadryas remain on the ground, or when feeding on a tree, climb down and run off. Early in the morning, before the troop moves out, young animals or subadult males sometimes climb acacias above their sleeping rock to rest or play. Males climb trees when looking for distant band or unit members, and juveniles escape onto thin limbs when chased by the heavy adult males who are unable to follow them there.

2. VARIATIONS IN PIGMENTATION OF SKIN AND HAIR

Elliot (1913) distinguished two species of hamadryas baboons. His *Papio hamadryas* lives in 'Abyssinia,' that is, in the westernmost part of our survey, and its mane runs from brown to ashy gray. The hamadryas of Southern Arabia, which are somewhat smaller but of the same color, are given the rank of a subspecies, *P. hamadryas arabicus*. In the Somali countries, a location between that of the two dark varieties, Elliot found a species of lighter color. He described it as *Papio brockmani* from a specimen obtained at Diredawa, in the eastern part of the present survey. The *brockmani* males are described as having conspicuous mantles of pale reddish brown to white hair.

As far as pigmentation is concerned, we found Elliot to be right. The western hamadryas are considerably darker than their counterparts at Erer-Gota and Diredawa. Still farther to the east, near Harar and Djidjiga, blond and, in adult males, white hues are predominant. Between the two extremes, however, there is a gradual transition. The characteristics of the two extremes are given below in Table III. The western variant may be roughly characterized as dark skinned and brown haired; the eastern as red skinned and yellow haired.

Table III. Pigmentation of baboons in the west (between Sendafa and Awash) and in the east (between Diredawa and Harar) of the survey area.

		West	East
Adult males	Face:	Dark grey with a shimmering of red	Bright vermillion
	Mantle:	Dark grey; lower back brown until adulthood	Light grey to whitish; some subadults completely bleached out; sides of head yellowish
	Anal region:	Pale red with a shimmering of grey	Bright light red
Adult females	Face:	Dark to blackish grey	Red with a shimmer of grey
	Coat:	Greyish-brown	Light brown to yellowish brown
	Anal region:	Dark grey; only the genital swellings are pink	Pale light red

3. VARIATIONS IN TROOP SIZE

Since the troop breaks up during the day, its size can only be determined at the sleeping rock. Table IV indicates that the sizes of troops, as well as the sizes of parties occasionally observed in open territory, increase greatly going from west to east. The five mean values derived from each of the samples in a given territory are significantly different ($F_{3;\,14} = 6.37$; $P < 0.01$).

Table IV. Changes in the mean size of troops and parties along the west-east axis between Addis Abeba and Harar.

Eastern Longitude	Area from which sample was taken	Number of troops or parties counted	Mean number of individuals in: troops at the sleeping rocks	parties encountered during the day
39° 10′	Aliltù near Sendafa	2	54	25
40° 9′	Awash Station	6	83	38
41° 1′	Afdem	1	82	—
41° 29′	Erer-Gota	5	110	90
41° 51′ to 42° 13′	Diredawa to Harar	5	354	227

The sharp increase in troop size between Awash Station and Harar is probably due to the fact that sleeping rocks become scarcer and food more abundant as one goes from west to east. Around Awash Station there is practically no agriculture, while the walls of the Awash valley offer almost too many sleeping rocks. A troop of only 14 animals was thus able to select a very satisfactory rock only two hundred meters away from another, larger troop. On the other hand, in the region near Diredawa, rich agriculture as well as the fruits of the opuntia cactus growing around the city are available to the animals. A party of baboons is seen stuffing themselves with chaff from the mill in the city nearly every day. But at night, up to 750 animals are forced to squeeze together on a steep river bank at the edge of the city; and even this does not provide enough room for all the animals, some having to sit on the level ground above. Thus, it looks as though the size of a troop is controlled less by sociological than by ecological factors: the geology and the food supply of the area cooperating to produce a west-east gradient (Fig. 9 and 10).

Fig. 9. The canyons of the area around Awash provide an abundance of suitable sleeping cliffs. Two troops sleeping in this valley numbered 14 and 47.

Fig. 10. 'Cone rock', an isolated rock formation in the plain north of Erer-Gota. The Cone rock troop averaged 120 baboons. Dark spots mark preferred sleeping ledges.

The smallest, completely isolated party which was encountered in open territory consisted of a one-male unit of six individuals; the smallest troop sleeping on a single rock consisted of two one-male units comprising a total of twelve individuals. The largest moving party was observed at midday and comprised 494 animals; the largest troop gathered at a sleeping rock was the one mentioned above, consisting of 750 animals.

4. VARIATIONS IN CLASS FREQUENCIES

17 out of 23 troops and parties were sampled to determine makeup by class (Table V). In these samples, either the entire troop or portions of a troop were recounted several times.

Table V. Mean percentage and standard deviation (in parentheses) of the main classes in the survey populations.

Number of individuals	Number of troops or parties	Adult males	Subadult males	Subadult and adult females	Young animals to 3 years old	Total %
2770	17	18.0	9.4	32.4	40.0	99.8
		(3.7)	(4.9)	(3.9)	(5.5)	

A west-east gradient in makeup by class was apparent in these counts: The percentage of males increased steadily and significantly ($P < 0.001$) from a value of 15.05% in Aliltù to a value of 22.6% near Harar. At the same time, there was a decrease in the proportion of females from 36.5% to 30.8%, while the proportion of young animals and sub-adult males appeared to remain the same. We have no explanation for this gradient.

5. VARIATIONS IN SOCIAL AND ESCAPE BEHAVIOR

The 3 months which we allowed ourselves for this survey, shortened by political events, were just sufficient to note the most basic characters of social organization in the samples taken along the west-east axis. The existence of at least some one-male units was well ascertained in 6 troops from the western through the eastern samples. Some one-male units walked alone at 30 to 300 meters from other baboons. In the extreme west, as in the east, neckbites (p. 36) were used by unit leaders when females moved away from them. The habit of adult males of waiting for the end of the column was frequently noted. Near Awash Station, a typical two-male team with a subadult follower was observed for two hours. In one of the troops at Aliltù, a unit leader performed the male presenting response, later termed notifying, in front of another who immediately followed him. In all resting troops, infants and juveniles formed play groups, some of them around a subadult male. The splitting up into definite bands was observed in 3 different troops immediately after departure from the cliff. The inconstancy of troop size was ascertained at Awash Station where a cliff was occupied by varying numbers of baboons in 3 different nights.

The only behavioral difference noted between western and eastern samples was a relatively higher frequency of male-male aggression in the large eastern troops. In fact, no such sequence was observed during the survey in the troops west of Diredawa.

The crucial question whether the following descriptions, beyond the region intensively studied, apply to a larger part of the Ethiopian hamadryas population can in one important point be answered with confidence: The one-male unit within the troop is an organizational feature found in samples throughout the population. It was found in Aliltù, in the small scattered parties of the arid bushland on the Awash river, as well as in the huge troops of the agricultural region near Diredawa, where parties entered the city.

6. THE BORDERLINE BETWEEN PAPIO ANUBIS AND PAPIO HAMADRYAS

The southern half of the western border of the investigated region broaches the area inhabited by the darkly pigmented anubis baboon (Fig. 1). In 1956 the boundary between the anubis and hamadryas areas lay at Wolenkiti in the Awash Valley (Starck and Frick, 1958). In 1960 we found a large party of anubis 70 kilometers farther east, near Metahara. Approximately half way between both places, we observed a party of eight baboons with one adult male. This male could be said to belong to neither the hamadryas nor to the anubis. The typical hamadryas mantle was missing on the male, his hair being of equal length down the entire back. His hair color, however, was characteristically hamadryas; the upper torso and arms were grey, while legs and lower body were brown. Moreover, the face was dark gray and the anal region a bright red. The females of the party could, except for their dark faces, not be distinguished from hamadryas females. Of course, it is possible that this male was merely a rare aberration, such as we did not again see in our 600 odd samples of adult hamadryas males. Another possibility is the following: The anubis are apparently advancing eastwards into hamadryas territory; the hamadryas, however, are not completely driven out, but have remained behind in enclaves at least and these hamadryas isolates have interbred with the anubis. Occasional interbreeding with anubis might be the cause for the dark pigmentation of the hamadryas in the west.

7. THE BORDERLINE BETWEEN THEROPITHECUS GELADA AND PAPIO HAMADRYAS

The northern half of the western border of our survey borders the region inhabited by the gelada. The Ghermanu River, which drains eastward into the lowlands occupied by the hamadryas, sends out a tributary some 50 kilometers northeast of Addis Abeba, in the neighborhood of Aliltù near Sendafa. The uppermost edges of the river's canyon rise 2500 meters above sea level, forming cliffs which are inhabited by geladas. In two out of three visits in December, we found small hamadryas troops and the permanent gelada population in the same canyon. Hamadryas and geladas joined to form mixed parties during the day, but at night separated to sleep 100 meters apart on the same cliff. On our second visit, we saw a hamadryas troop migrating down the valley towards the plain and on our third visit, we found only geladas. The behavior of the two species (p. 173) indicated that the hamadryas were superior to the geladas, which outnumbered them by far, even though no interspecific fights were observed. I suspect that during the dry season, individual hamadryas parties migrate out of the eastern lowlands up into the canyons of the river to settle for a short time in gelada territory which is more humid and rich in grain. They probably return to the lowlands during the rainy season when the region becomes quite cold because of its altitude.

III. THE BROAD SAMPLE

In February 1961, we made a choice among the samples covered by the initial survey. For the broad sample we looked for an undisturbed population which seemed to be representative in as many variables as possible. For practical purposes of observation, low flight distances, sparse vegetation and a hilly ground relief were preferable. These conditions are met by the region between Erer-Gota and Garbelucu (Fig. 11). The hamadryas population of this region lives 190 kilometers away from the nearest (western) edge of the species' area. The mean troop size was determined again and found to be 10 individuals above the value found during the survey. The degree of pigmentation comes close to the average, although it tends slightly towards the eastern extreme.

The village of Erer-Gota lies about 1200 meters above sea level at 9° 32′ N. Lat., 41° 23′ E. Long. at the northern foot of the Ahmar

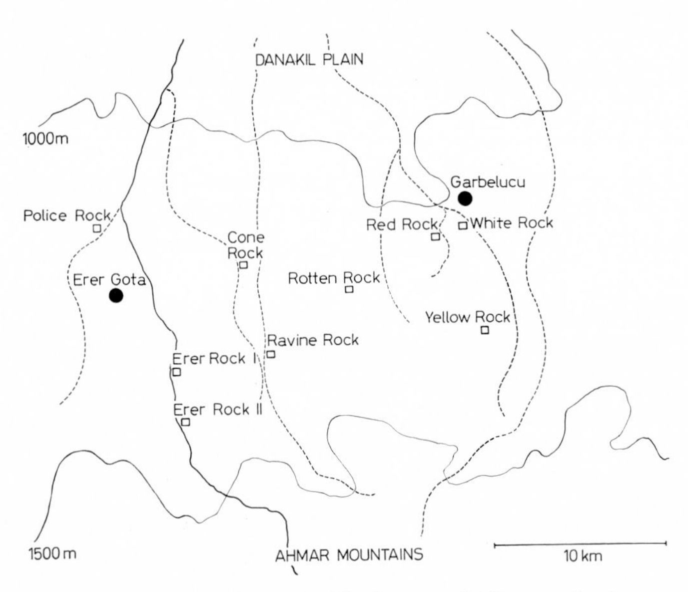

Fig. 11. The area of the broad sample. The heavy solid line marks the permanent Erer river; broken lines represent usually dry river beds. Black circles are villages. The sleeping rocks are indicated by squares.

Fig. 12. A typical section of the Erer-Gota habitat seen from the north, with the foothills of the Ahmar mountains in the background. The Red Rock (right center) rises above a dry river bed.

Mountains. These mountains extend from West to East and form the southern border of the arid Danakil Plain. Their slopes are covered with forests and woods. Before their rivers flow into the Danakil desert to dry up, they irrigate a hilly strip about 20 kilometers wide which stretches out in an east-west direction along the foot of the mountains. This strip is the habitat of our broad sample (Fig. 12). It is thinly wooded with acacias about 4 meters high, which stand 5 to 10 meters apart. Large trees are only found on the banks of the few permanent rivers. To the north the acacia strip rapidly changes into open grassland and semidesert. Sleeping rocks are more numerous near the mountains than in the northern plain. No cultivated fields are found in the northern part of the strip, and at the southern edge, near the foot of the mountains, only a few such fields are situated on the banks of the rivers. Only here are the animals occasionally chased away. The northern troops were never seen in the vicinity of such fields, nor were they ever molested by humans. Lions, leopards and cheetahs, though rare, were observed now and then; but they were never seen in an actual encounter with the hamadryas. Observations of that sort can only be made in reservations where predators are not kept away by the presence of the observers.

We were usually able to see the animals through the relatively sparse vegetation from a distance of 10 to 80 meters. The profile of the terrain itself played an important part in allowing us to make our observations; since the hilltops were usually somewhat farther from each other than the baboons were from us, we could observe the animals on a slope while we were on the opposite ridge. More details on the ecology of this area may be found on page 157.

1. TROOP SIZE AND TROOP COMPOSITION

Our second troop counts near Erer-Gota are summarized in Table VI. The 'regularly used' rocks were occupied by baboons every night. They were situated in the northern, flat part of the sample area, where rocks were relatively sparse, and they attracted large troops. In contrast, many of the apparently perfect rocks in the foothills of the Ahmar mountains were only occasionally used by small troops. This again suggests that the density of rocks affects troop size.

Table VI. Size of troops in the Erer-Gota population as determined from counts taken at four regularly used and one rarely used sleeping rock.

	Number of counts	Mean number of individuals	Largest number	Smallest number
Regularly used rocks	24	120.8	156	62
Irregularly used rock	21	47.7	96	12

Five counts of entire troops (totaling 440 animals) and 36 counts of portions of troops (totaling 911 animals) form the basis of our calculation of the frequencies of the various sex-age classes. The data were obtained as the animals crossed an open area in long, drawn out columns. Under these conditions, the sex of all adult and subadult members could be determined, but the sex of infants and juveniles could usually be determined in only one half of the animals. The following results are published only because I do not see any other technique of counting possible, unless, of course, entire troops are caught and marked, or killed.

In five smaller troops, it was possible to record the sexes and ages of all juveniles and infants. The total for these counts coincidently came to 100 animals. The ratio was 46 males to 54 females. In the incomplete counts described above, we were able to identify the sex of 207 juvenile animals: 44.3% were males; 55.7% were females. Since these data correspond fairly well, we may conclude that in the incomplete counts, both sexes were identified with equal frequency, and that the true ratio can be derived from the observed ratio. In Table VII all such calculated figures are in brackets. Percentages without brackets are the results of complete classifications.

The quantitative data on class proximity (p. 88) were used to calculate the class frequencies from another, independent set of data. The new figures deviated not more than 0.5% from the figures of Table VII, with the exception of the one-year old males whose percentage appeared as 5.2% from the proximity equations as compared with 6.5% in Table VII.

Table VII. Percentage of age and sex classes in the population at Erer-Gota. Total size of samples: 1351 animals. Bracketed figures are calculated as described in the text.

	Adult	Subadult	3½–2½ years	2½–1½ years	1½–1 years	½–0 years	Total sex classes
Males	22.9	6.0	(4.3)	(4.9)	(6.5)	(1.9)	(46.5)
Females	32.6		(5.6)	(6.1)	(7.5)	(1.8)	(53.6)
Total age classes	61.5		9.9	11.0	14.0	3.7	100.1

From Tables V and VII it is apparent that 60% of the populations consist of adults and subadults, and that there are almost as many males as females. In Kenya's anubis groups, although about half the population are also adults, there are two or more adult females for every adult male (DeVore, 1962). In the chacma groups studied by Hall (1962 a), the ratio is four or more females to every male. From our own observations, the gelada troops near Aliltù contain only 5 to 15% adult males. Our hamadryas population thus included an unusually high percentage of adult males, as compared with other baboon species.

Since only one solitary individual (an adult male) was ever observed, the tertiary sex ratio of the population can be calculated directly from counts of the troops themselves. With the inclusion of subadults, this ratio comes to 1:1.13 in the broad sample. The socionomic sex ratio must be based on the units of reproduction, i.e., on the one-male units. In 68 one-male units counted without sampling error, 1.86 subadult and adult females were found for every adult male unit leader. The sex ratio for young animals up to 3.5 years of age is 1:1.17 and is thus indistinguishable from that of subadults and adults. Only among black infants may males equal the females: In complete counts of infants of entire troops, there were 13 black males to 11 black females. The low figures for black infants are explained by the fact that the counts were made three months before the peak of births (p. 177).

We found no evidence for the argument that adult male baboons reduce their number by fights among themselves. The most serious battle wounds we ever came across in our observations of several hundred fights between adult males and several fights between bands, were a torn-open nostril and a lightly bleeding shoulder. The fighting technique (p. 47) tends to preclude the infliction of serious injury.

2. POPULATION DENSITY

Counts of all of the animals spending their nights on six sleeping rocks between Erer-Gota and Garbelucu (excluding the three westernmost rocks, see Fig. 11) were accomplished in two nights, one week apart. In making these counts, each of us would cover three rocks, either as the animals arrived in the evening or departed in the morning. On our first count, we found a total of 446 animals; on the second, 452, whereby the same two rocks were found unoccupied both times. The counts are probably correct within $\pm$ 20 animals. On the map these six rocks fit within an area of 250 km². Accordingly, the population density within the area was 1.8 baboons per square kilometer. The density of savanna baboons in the areas studied by Hall and DeVore (1965) is nearly three times as high.

IV. ORGANIZATION AND LIFE CYCLE OF THE ONE-MALE UNIT

In comparing the desert baboon with the savanna baboons (KUMMER, in press), one finds little difference in the gestures of social behavior, but a striking contrast in social organization. There is only one well defined level of organization in savanna baboons, the stable *group* of several dozen individuals. Within the group there are no constant subgroups into which the group will occasionally split, nor does the group join with neighboring groups. In the hamadryas, both internal organization and relations with neighboring groups are changed. The band, which in many respects resembles the savanna baboon group, may split into one-male units which may forage independently during the day. On the other hand, several bands may join each other in using the same sleeping rock, thus forming a troop. The hamadryas organization therefore has three levels: the troop, the band, and the one-male unit. In describing these levels we will start with the lowest, the one-male unit.

1. COMPOSITION OF THE ONE-MALE UNITS

First, the one-male units of the Erer-Gota population shall be considered statistically on the basis of the broad sample. In our preliminary report on the close study (1963), we have shown that the identified one-male units met the following criteria: (1) The members of a unit stayed together more closely than they stayed with the rest of the troop. (2) The frequency of social behavior was significantly higher among the unit members than between members and outsiders. (3) Both of these characteristics were retained throughout the months of observation. The available time of course did not

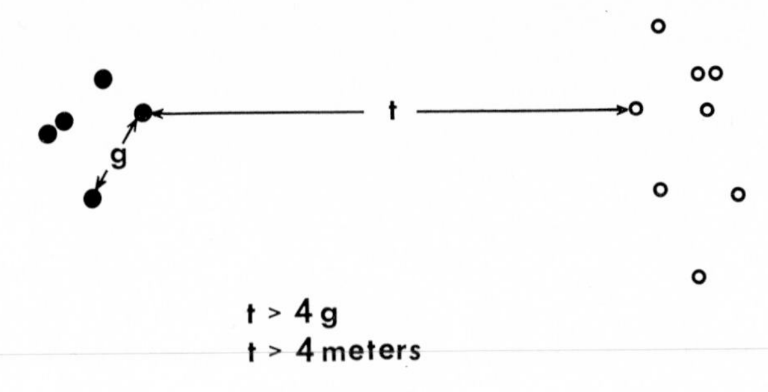

Fig. 13. Criterion of spatial detachment used in identifying 'detached parties' (see text). Different symbols represent members of different parties.

allow us to investigate the broad sample in such detail. We therefore studied its social units only on the basis of the first criterion, spatial coherence. A set of animals was recorded as a 'detached party' if at any given instant, the smallest distance (t) between the set and another animal was greater than 4 meters and 4 times larger than the greatest distance (g) between two neighboring animals within the set (Fig. 13). This arbitrary criterion represents an unusually high degree of spatial detachment within a hamadryas troop. With reasonable certainty, the members of such an isolate temporarily follow each other, and communicate with each other more often than with any other baboon. The permanence of social units defined in this way can, of course, vary greatly. The method records parties which might break apart at any moment, such as young animals playing together or subadult males grooming each other. The close studies of identified individuals have shown that of all types of detached parties, only the one-male unit stays together continually; none of the 6 units studied from 1 to 6 months was seen to change its leader, and only one female disappeared. The other types of detached parties found with this method are described on page 86.

Figure 14 shows the compositions of detached parties including one, and only one, adult male. Since we were interested in the relative frequency of the different types of one-male units, we only included complete counts of all one-male units within a band or troop. Such counts were successful only during staggered entries onto a sleeping rock. They showed that a one-male unit includes an average of about 2.3 females. In another, non-random sample of 163 units which were found separated from more compact parties, this figure amounted to 2.1 females. Only the large units with 7 and more

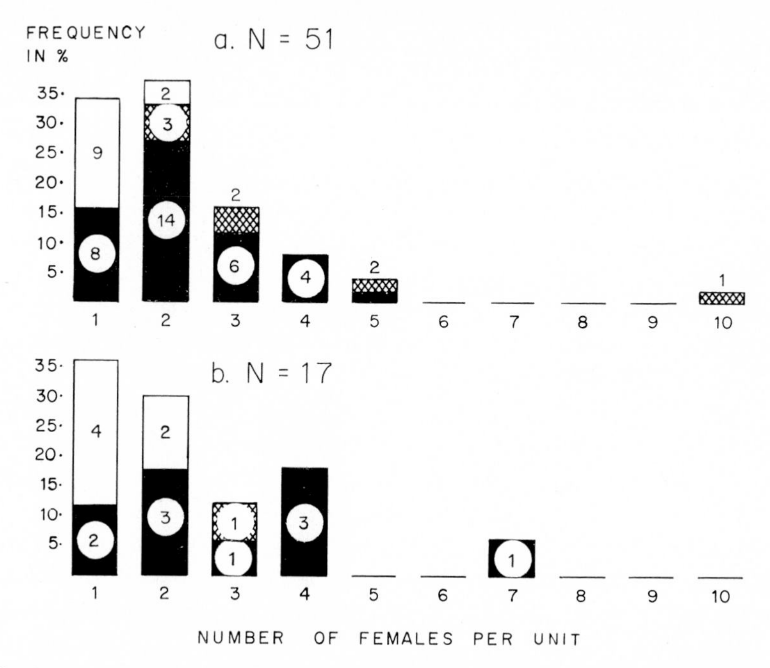

Fig. 14. Frequencies of one-male units containing varying numbers of females. (a) Totaled counts of parties of the broad sample. (b) Entire L-band of White Rock troop (close study). Black: Units in which all females are adults and subadults; white: Units in which all females are juveniles; dotted: Mixed units with both adult and juvenile females. Figures in columns give the absolute frequency of each unit type.

females were not fully represented in the non-random sample. The close study later revealed that such units contain many mothers of black infants and that they tend to remain in compact aggregations.

It may at first seem unwarranted that the counts also include immature females. The close study will show, however, that females of about $1\frac{1}{2}$ years will often, without their mothers, follow an adult male's lead, behaving like an adult female consort being led by her male. Therefore, each frequency column in Figure 14 is divided into a maximum of three divisions whose heights indicate the frequency of purely adult, mixed, and purely juvenile female membership. On the average, a one-male unit of this sample contained 1.86 subadult and adult females and 0.44 females $1\frac{1}{2}$ to 3 years old. On the basis of Table VII, one may estimate that more than half of the $1\frac{1}{2}$ to 3 year-old females already followed an adult male rather than their mother. In the close study, all females older than 2 years were found to follow a unit leader, not their mother. The juvenile females were found significantly more often in small than in large units. Thirteen out of the 30 juvenile females even were the only consorts of young adult males. The close study will describe these associations as 'initial units'.

In the close study, all sexually mature females were found to live in one-male units, whereas some males of all age classes lived outside such units, although still within the troop. A comparison of the socionomic sex ratio in the 68 units of Figure 14 with the tertiary sex ratio of the population shows that about 23% of the adult males in the sample of Figure 14 had no females at all.

Some one-male units are accompanied by one or more subadult males as 'followers', which are groomed now and then by the adult females of the unit, but seldom copulate with them. In the samples of Figure 14, 36% of the units having adult females also had male followers, but none of the 17 units having only juvenile females was accompanied by followers. The reason for this was found in the close study (see p. 54). The number of adult females in a unit, however, had no relation to its ability to attract followers.

2. CLOSE STUDY OF ONE-MALE UNITS IN THEIR ZENITH

The bare numerical framework can now be amplified in a few critical areas by data from the close study. At the same time, methodical emphasis will switch from spatial arrangement to the pattern of social interactions. Six months after the beginning of our field work, I selected 6 one-male units of the White Rock troop and 2 of the Ravine Rock troop. The 8 units were selected as representing the essential types of the compositional range. They included units with leaders of various ages, units having only juvenile females, and one of the extremely large 'mothers' units.' Besides the long range observations of the identified units, we shall use several hundred short records of non-identified units belonging to the same compositional types.

Each unit was observed at intervals for several weeks or months, on the sleeping rock and in the waiting areas. On route, the identified units could be observed only when they split off from the others. At an escape distance of 40 meters, it is rarely possible to get into the middle of the troop. The following data, therefore, as well as the

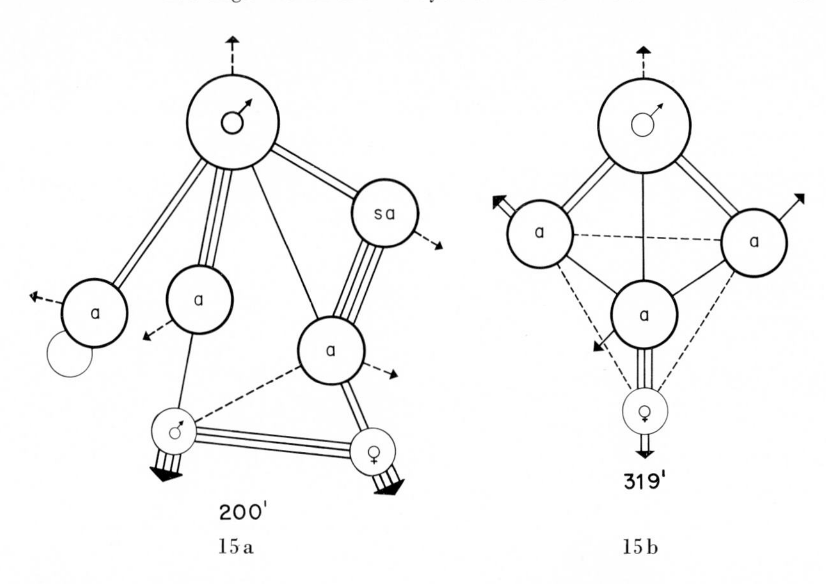

Fig. 15. Sociograms of one-male units at the peak of their development (Circum and Smoke). (a) Circum (male-female distances are 69, 43, 76 and 50 cm); (b) Smoke (male-female distances are 31, 46 and 36 cm). Large circle = adult male; medium circle = female consort; small circle on the radius of a female = her 6 to 18-months-old child; attached circle = her black infant. a = adult female; sa = subadult female; 3 = 3-year-old female; 2 = 2-year-old female. Connecting lines express the percentage of observation minutes that contained an exchange of the gestures specified on p. 180. Arrows indicate exchanges with animals outside the unit: ---- less than 3%; —— 4 to 10%; ═══ 11 to 20%; ≡≡≡ 21 to 30%, etc. The graph distance between the leader and each female is proportional to the mean of the recorded distances. The total time of observation is indicated in minutes.

sociograms, incorporate only the behavior near the sleeping rocks. Fortunately, this is where all sexual behavior, fighting and most grooming and play occur. When the troop is on its daily route, social behavior consists almost entirely of the subtle interactions of spacing, while obvious social contacts can disappear altogether for hours. The spacing of the moving unit and the behavior correlated with it will be described in their wider context, in the chapter on mechanisms of troop movement.

I shall first describe the social life of one-male units at the zenith of their development when their reproductive rate is highest and their internal and external activities fully developed and balanced. Accounts of the other unit types will then follow in the order which reflects the probable history of a unit leader's life.

At their peaks, hamadryas units contain from two to five females, most of which are adults. Both 'Smoke' and 'Circum' (Fig. 15) were fully grown males in their early prime. Their coats had not yet taken on the silver sheen of older males. Smoke was the leader of one subadult and two adult females, one of which had a 1½-year-old female child. Circum had three adult females and one which was not quite subadult. His adult females had one child each: a one-year-old male, a one-year-old female and an infant which had just been born before the start of my observations.

a. Leadership and Cohesiveness

The members of a one-male unit always stay together within the troop (Fig. 16), but the unit rarely leaves the troop to go off by itself. On only two occasions did we encounter completely isolated one-male units of 6 and 8 baboons foraging in the savanna. On the march, leaders frequently look back at their females, and they respond with threats when the females lag behind or get too far away. The

Fig. 16. One-male units cuddling together on a cool morning. The subadult follower on the left sits somewhat apart from his unit. Although units avoid mixing, they rarely detach themselves to this degree when the troop is united.

17a

17b

Fig. 17. (a) A young leader (Guy, p. 71) stares and raises brows at his straying female. (b) She comes back and presents to him.

mildest threat is the stare with raised brows (Fig. 17). As the threat increases in intensity, its form becomes more severe:[1]

A fight breaks out on the sleeping rock. As soon as it begins, Smoke looks up, advances quickly to the farthest of his females and hits her gently on the head with his hand.

One of Smoke's females squeezes herself among the other females, moving away from him. At this, Smoke raises his brows and stares at her. Though she has not noticed the threat, he does not continue it.

A male, during the daily march, looks back for one of his females in oestrus. As she appears from behind a small ridge he lunges at her. Uttering a staccato cough, she runs toward him.

A male, having just arrived at the sleeping rock, turns suddenly and rushes 30 meters back along the on-coming column. An adult female from the farthest party runs toward him and receives a bite on the back of the neck. Squealing, she follows the male up to the sleeping rock where his other females are waiting.

The bite on the nape of the neck or on the back is the sharpest reaction a unit leader makes to a straying female (Fig. 18). It immediately causes the female to follow him closely. Sometimes the female is lifted off the ground during the biting. In one case, a female

Fig. 18. A mother with an infant screams as her leader bites her on the back.

[1] Small print is used for records of actual sequences. The reader who is interested only in the conclusions can skip these sections.

fell five meters down the side of the sleeping rock during a neck bite. Such brutal proceedings, however, are not very frequent and the bite on the neck hardly ever produces a bleeding wound. On the average, no more than one bite per day is given by a leader. When a unit leader's possession of a female is threatened by other males, he puts his arms around her or covers her with his body in a hunched, sitting position (Fig. 41 b), all the while threatening his opponents. This behavior is only seen during widespread aggression among the males of a troop. Females respond to a mild threat by running to their males uttering a staccato cough. When bitten on the neck or back, they respond by screaming and eventually by pressing themselves to the ground in a motionless crouch.

A leader does not allow equal freedom of withdrawal to all of his females. Females with new-born infants withdraw from active unit life during the first weeks and are often allowed to go farther away from their males. Even a distance of 40 meters does not necessarily provoke an attack.

A mother, together with her newborn child, was seen as she sat down 5 meters away from her leader. Later, another female, one in oestrus, came to her and sat down beside her. She, however, was soon brought back by the male with a neck-bite. With this, the mother also returned.

In general, juvenile females are allowed more freedom of movement than adult females, except where the juvenile is her male's only female.

A one-male unit is kept together principally because of the potential threat from its leader. When left to themselves, or lost by the male, females will sometimes remain out of sight of their unit for as long as half an hour. Apparently, however, females are often aware of the consequences of their behavior:

During the midday rest at a watering place, a male comes along the river bed. An adult female which he has apparently overlooked, rushes out of the brush, presents her rear to him while still a meter away and proceeds to walk back ahead of him to where he came from, staccato coughing softly.

As a large troop is breaking up into three separate and compact bands of about 100 each, a female rushes back out of the first band to the second, over a distance of 60 meters. There she is greeted by a male with a bite on the neck.

On the other hand, there are some situations in which the female actively seeks the proximity of her male. For example, during a heavy rain, females may creep in close to their male while he turns his back to the wind. During aggression between two unit leaders, females try to get as close to their own male as possible; though during actual fighting, the females will stay up to 10 meters away and rejoin their leader after the fight, sitting in a centrifugal row or clump in the leader's 'shadow of attack' (Fig. 41 b and c).

While on route, a female jumps away from some object in the grass and immediately runs to her leader some meters away.

If a female happens to be attacked by another animal, she will sit down very close to her leader who, in turn, will answer the stranger's threat. Females will also join their leader if another male approaches too closely. Generally, the female seeks a position which places the male between her and an outside threat, and the male will sometimes assist her in the attempt. When a unit withdraws from an observer, the male, who ordinarily leads the way, often waits for his females to catch up and then to precede him, while he brings up the rear.

Units rarely intermingle. A female separated from her leader risks a neck bite, whether she is separated from him by excessive physical distance or by an outsider who sits between her and the leader. Relative, social distance can thus substitute physical distance. In the following sequence, the females seem to anticipate their leader's reaction:

One evening Circum is sitting some distance away from the sleeping rock, grooming one of his females. His other three females are already sitting on the cliff. A neighboring leader, together with his own females, has to climb through Circum's females in order to get to his own sleeping ledge. At this, Circum's females arouse and run to Circum screaming. Circum immediately bites one of them on the neck and places himself bellywards on top of all three. After this, the three females begin to groom Circum frantically. The following evening, although Circum is now sitting with his unit, precisely the same thing happens. All this despite the fact that this neighbor was certainly well known to Circum's females and both groups had formed a small independent troop for several days.

On six occasions, we observed a leader looking for a lost female. On one of these occasions, the leader was known to us and the female definitely belonged to him. The following is an illustration of the typical behavior of males when one of their females has moved out of sight:

A 2½ year-old female is seen lagging 100 meters behind her party. Suddenly a young adult male runs out of that party back through the grass. Before he gets to her, the female darts away from the direction of the march. At the place where she has turned off into the brush, he pauses, stands up on his legs and scans the surroundings for her. Spotting her, he runs out in pursuit. She, in turn, races forward in the direction of the march and disappears for good. Again he gets up on his legs, then runs to one of the older males in his party who happens to be looking on. The young male starts to produce liquid feces, while running back and forth, accompanied by a steady 'Bahu' bark. Here and there he races up a tree and scans the area. Again he is seen running to the older male and looking at his face (cf. p. 128). During the course of his continued searchings, which we observe for 12 minutes, we lose sight of him.

WASHBURN and DEVORE (1962) have reported that among Kenya's anubis baboons, sick animals are left behind as soon as they are unable to follow the rest. This is also true for *troop* members among the hamadryas:

A senile or sick male whose coat is sparse and ragged is seen to arrive alone at the sleeping rock 20 minutes after the rest of the troop. Observations on the next day show that he has no females.

The members of a *unit*, however, will not easily leave each other in the lurch:

Seven minutes after the entrance of the last party onto the sleeping rock, an adult male together with a single female carrying a black infant on her belly approaches the cliff. The infant does not hold onto its mother so that she is forced to carry it on her arm and to hobble along on three legs. The male in turn keeps stopping now and then so that the female can catch up to him. On the rock itself, the infant lays motionless in his mother's arms.

In summary, the cohesion of a one-male unit at its peak is primarily enforced by the aggressive herding behavior of its leader, since the females, for their part, will stay away from their unit as long as they are surrounded by other animals. Females demonstrate a centripetal tendency which binds them to the troop, while adult males leave the troop more readily. Nevertheless, if a leader leaves the troop, his females will follow him, not the troop. Additional insight into the nature of unit cohesion will be obtained when we deal with the initial units and the transplantation experiments.

b. Sexual Relationships

True copulation among hamadryas consists of a series of mountings occurring at intervals of three to eight minutes. Ejaculation occurs only after several such mountings. True copulation is further marked by the male's feet grasping the female's ankles. It occurs only when females have a perineal swelling, which is the visible correlate of oestrus (ZUCKERMAN and PARKES, 1930). There is evidence (p. 178) that females of the same unit tend to synchronize their menstrual cycles.

Individual mountings can be initiated by either partner. If the female takes the initiative, she presents her swelling to the male, as in Fig. 17 b. The male may initiate mounting by first looking at the sitting female's swelling and touching it lightly with his hand; or by grasping her by the hair on her head and pulling her to a standing position. After dismounting, the pair engage in intensive grooming. It appears as though the male grooms more often before ejaculation than after, although this difference is not significant. The frequency of mountings with intromission per 100 animals varies between 0.7 and 12.2 times per hour, depending on the time of year (Fig. 68, p. 177). These figures refer only to behavior in the morning and in the evening at the sleeping rock. Copulation was not observed during the daily march.

When the hamadryas colony in the Zurich Zoo contained 7 sexually mature females and only one old male, females in oestrus often mounted other females (KUMMER, 1957). In the wild this behavior was observed only once:

A two-year-old female in oestrus mounts an immature 1½-year-old female. After this, she approaches her unit leader and presents to him so long until he finally is induced to copulate with her.

During a copulation series, a female was also observed to mount a male:

After his first intromission, a male walks away from his consort and then looks back after a few paces. Thereupon, the female follows him and presents. After a second intromission, the male again withdraws. The female again follows and repeats her presentation. This time, the male turns away. With this, the female mounts him, then jumps upon his back, where standing on all fours she is carried along briefly.

Unit leaders sometimes mount their females in anoestrus. In these cases the male's feet often remain on the ground, and intromission, which is difficult with females not in swelling, was observed only once. Such mountings of a female in anoestrus were limited to times of tension, as for example, when the neighbors of the pair chased each other, or when a leader was hesitant about entering onto a sleeping rock strewn with unfamiliar food. The mounting of a female during anoestrus also occurs during threat (raising brows and pumping cheeks) between unit leaders (Fig. 41 c). Juvenile males often threatened us while mounting each other. It is possible that mounting during tension has some motivational relations to clinging.

Adult males copulate only with females belonging to their own unit. We have never found any exception to this rule. At most, a male would follow the swellings of a passing female with his eyes. This means that males with only one female may sometimes not arrive at copulation for months at a time. Those having no females at all may remain without copulating for years. The first case is illustrated by the male 'Fork' whose only female was first pregnant and then nursing during the entire six months of observation. Occasionally, males attempted to aquire females from other units during prolonged aggression in the troop, but none of these females was in oestrus at the time. On the other hand, we have observed an adult male withdraw when he was approached by another male's female in oestrus. The adult males seem more concerned with keeping a set of females following them at all times than with achieving a chance copulation with a strange female. This conclusion will be supported by the observations on initial units, whose females are sexually immature, by the fights over anoestrous females, and by the transplantation experiments.

Females, on the other hand, will copulate not only with their unit leader, but also with subadult and juvenile males. Since a long series of mountings is almost impossible in the vicinity of the unit leader and since many of these young males are still sexually immature, such copulations may seldom result in pregnancy.

Typical sexual interactions of *adult* females with outsiders may look like this:

An adult male is leaving the sleeping rock followed by his females. On the way, one of these presents her swelling to a subadult male who then mounts her. As they are copulating, the unit leader looks back, causing the two to separate and the young male to flee. The leader attacks his female and after biting her on the back, mounts her himself.

A young subadult male is seen copulating with an adult female in oestrus behind the back of her leader who walks in front. Another unit leader, who happens to be passing, briefly looks at them, but does not interfere with the copulation. Seconds later, the young male runs over to the owner of the female, presents to him and is mounted briefly.

A three-year-old male is observed repeatedly copulating with an adult female in oestrus. After the first and second time, she turns to look at her leader who is facing away from them, and then grooms the young male. After the third time, she runs to the unit leader and presents to him. Then she continues to groom the young male.

Adult females thus attempt to copulate with young males in the immediate neighborhood but behind the backs of their leaders.

In many instances they provoke such mountings by presenting to the young males. We have regularly seen a unit leader attack the female and not the male when the couple happens to be caught. Other unit leaders do not interfere with the young males' attempts. After the copulation, the female tends to approach her leader and to present to him, even if he has not noticed her transgression. Copulation with outsiders was especially frequent in June when a large percentage of the troop's females were in oestrus.

The *juvenile* females' sexual contacts with outsiders differ from those of adult females. They get their first monthly swellings at the age of about two years and by then are already members of a one-male unit. If the unit is at its zenith, they share their membership with several adult females. Since the latter are preferred by the leader, these juvenile females rarely copulate with him. However, the little attention paid them by their leader allows them to leave the unit for short periods. They are then seen to move freely through the troop, sometimes playing in groups of juveniles. During these excursions, if in oestrus, they will often approach a 1½–2-year-old male and engage in a technically correct 'copulation' series with him, usually out of the unit leader's sight and at least 10 meters away from him. Often, the juvenile females will run back to their leader after each mounting, present to him or sit with him for a moment. They then return to their young partners who are usually found waiting behind some projection in the cliff, and continue copulation. Juvenile hamadryas baboons thus simulate the temporal consort pair relations typical of adult savanna baboons.

A 2½-year-old male is sitting alone. A three-year-old female in oestrus is standing 5 meters away with her one-male unit, which contains two adult females. The juvenile male exchanges glances with the three-year-old female, then raises his brows (mild threat), whereupon the female comes running to him and presents herself. Intromission occurs after a first unsuccessful attempt. After that she quickly returns to her unit and sits down very close to her leader who apparently noticed nothing. The juvenile male follows her and sits down next to her, watching as the leader is groomed by one of his adult females. Later, he moves close to the leader and is immediately threatened by the grooming female. Screaming, he clings on to the back of the leader. She continues to threaten him for half an hour whereupon he withdraws.

Almost identical scenes were observed in the zoo colony. Our quantitative data on age-sex classes of neighbors (p. 88) will support the observation that adult females attract mainly young subadult and three year-old males, while juvenile females mostly attract one and two year-old males.

Obviously the hamadryas male goes through two separate phases of sexual activity during the course of his lifetime. The first phase occurs while he is still a juvenile or early subadult. Despite his sexual immaturity, he engages in 'copulation' with females in oestrus. This phase is followed by a long period of sexual inactivity during his late subadult and early adult years. Finally, the fertile second phase sets in, during which he only copulates with the females of his own unit. Females, on the other hand, having once reached sexual maturity, probably copulate during each oestrus throughout their lives. Beginning with 'copulations' with juvenile males outside their unit, they go on, in subadulthood, to mate mainly with their own leader.

As to the wider social effects of oestrus, our comparative study (1965) shows that females in oestrus, both in captivity and in the wild, groom less and are less often groomed than females in anoestrus. These differences are statistically significant. In part, sexual behavior may replace grooming in its function of strengthening of the relationship with the male at this time. The general reduction of social activity and the changed neighborhood of the female in oestrus are shown in Figures 39 and 40. During oestrus, juvenile as well as adult females stay farther away from their neighbors, and the usual tendency to stay within the troop is not observed. Thus, units in which all of the females are in oestrus tend to keep somewhat apart from the troop.

In many cases, a female in oestrus will not follow her male; instead, the male is often seen to follow her. In his attempts to keep her close, he may substitute the usual threats by grooming her.

A unit leader, together with his four females, is leaving a watering place where they have stopped to rest. A subadult female belonging to the unit and in full oestrus simply remains seated. After going no more than four meters, the leader stops, looks around at his female for some 10 seconds and then comes back to her. Two minutes later he again tries a move in the same direction; again all his females follow him with the exception of the one in oestrus. Within 30 seconds the entire unit returns a second time. The leader begins to groom his reluctant female. The grooming lasts about a minute and after eight minutes another move is attempted. Again, the female does not follow but remains seated scratching herself. Once more, the leader returns to his female and begins to groom her for another three minutes. Seventeen minutes later the leader tries a third move, but with no more success than before. This time the unit leader returns immediately to his female and sits down watching her. She, in turn, turns her back to him and proceeds to scratch herself.

The tolerance of a leader towards a female in oestrus is often extraordinary. In the chapter on troop movement, we shall see how a female in oestrus belonging to an influential leader was able to stop the movement of a large advanced party simply by straying away from it. As she left, she was followed by her leader, who in turn was followed by other units until finally the entire party was moving in her direction.

c. Grooming Relations Among the Adults

Of all social activity, mutual grooming takes up the largest amount of an adult baboon's time. At the sleeping rock adult males were grooming in 12% of the observation minutes and were being groomed in 19%, whereas sexual and aggressive interactions taken together occurred in only 1.5% of the observation minutes (KUMMER and KURT, 1965). Adult females in anoestrus groomed in 28%, were groomed in 21%, and had aggressive and sexual encounters in 0.9% of the minutes. Thus, the sociograms of the identified units are to a large extent sociograms of social grooming. Whereas sexual behavior and fighting were observed only at the sleeping rock, grooming was also observed on occasional stops made along the daily route.

The grooming response can be released when an animal places himself on all fours in front of his partner, lowers his head and raises his tail in a gentle arch, thus presenting his side. Presenting of the rear can also release grooming behavior. Sometimes one partner pulls the other to him by the hair and begins to groom it. Grooming by females is primarily concentrated on the unit leader and their own infants, very little being done among themselves. Only in the sociogram for Circum's unit, two females are seen to have a relatively large amount of interactions with one another. It is possible that since one of them is barely subadult, they were mother and daughter, a relation also found in Fork's unit (Fig. 36). Apparently, grooming between two females of a one-male unit and even more, grooming between a female and a male follower, come very close to overstepping the borderline of a leader's tolerance. Before such grooming occurs, the animals in question often exchange glances and then back several paces away from their unit leader.

From the figures mentioned above, it is evident that unit leaders assume the passive role more frequently than do adult females. This difference is statistically significant. Smoke, for example, was groomed by his three females five times as often as he groomed them. In a troop of about 50 animals, the following typical count was taken: Among adult members, 7 females were grooming males; 1 male was grooming a female; 2 females were grooming other females and 1 male was grooming another male. Juvenile females eagerly strive to groom their leader but are successful in doing so only when no adult female is grooming him. Moreover, should a juvenile female try to get her leader to reciprocate by presenting her flank to him, she seldom meets with success.

Grooming within a one-male unit is subject both to competition among its females and to the preferences of the male. In each of Circum's and Smoke's units, there was a female who was able to displace the other females from grooming the leader. In return, each of the two favorites was groomed by her male for more than 1/3 of the time spent by the pair in grooming, whereas the other females enjoyed the passive part in only 1/10 of the time. Circum's favorite female was usually nearest to him and would most often be the first to follow. Circum showed a grooming preference for yet another of his females, who had a very young infant and never approached him of her own accord. There may be two female ranking orders which overlap each other; an autonomous one among the females and one dependent upon the leader. Oestrus alone is not sufficient to make a female a favorite.

On the surface, it often appears as if grooming were used to strengthen a social bond when it is in danger of breaking apart. In the zoo group an old male was never seen to groom his females as long as he had seven of them. But as soon as he had lost all but two of these, he was frequently observed grooming them. Again, the young males in the zoo colony were often seen grooming their first acquired females, but they groomed less and less frequently as their units got bigger. The above interpretation is also suggested by short grooming bouts which occur in a number of special contexts. For example, females will groom their males after having received neck bites or after aggressive encounters with other females. At such times grooming movements are much more rapid than usual, and the normal removing and eating of particles from the skin is left out. We have also observed rapid and intense grooming in the following instances:

A lost infant is found again by its mother and groomed at high speed.

A female hesitatingly approaches a water hole in spite of our presence. Then she runs back to her male and begins to groom him vigorously.

The leader 'Fork' has made way for another male to pass him on the sleeping ledge and finds this male a few minutes later staring at him and his (Fork's) female. Immediately, Fork turns to the female and begins to groom her with high speed.

d. Aggressive Behavior Among Females

Aggression by the male against his females is produced by anything that reduces the behavioral and spatial integrity of his unit. Aggression among females is not so clearly related to certain external events. It is always carried out in front of the male whose support the females try to enlist. Therefore, the issues of dominance among the females and of competition for the relations with the leader are not easily distinguished. No aggression between females of different units was observed in the wild. When it happened in the zoo colony, the opposing females were brought back to their units by neck bites from their leaders.

Should two females simultaneously attempt to groom their leader they soon begin to scream at one another across his back, and hit out at each other until one of them withdraws.

An adult female and a subadult female belonging to the same unit begin to scream at each other for an unknown reason. After chasing the subadult female away, by hitting out at her, the adult female proceeds to rapidly groom her leader. Thereupon the subadult female comes back and begins to groom the leader from the other side. Very few of the animals on the sleeping rock are engaged in grooming and most of them are asleep. The two females, however, continue to scream at each other over the shoulders of their male. Finally, the subadult clings onto the back of the male. At this, the male bites the adult female on the back of the neck.

In the Zurich Zoo colony, such situations usually led to the so-called 'protected threat'. One female would threaten another by raising her brows and slapping the ground, while at the same time presenting to her leader and thus preventing his attack on her. This often resulted in the leader's attack on the threatened female. The protected threat was frequently observed in the zoo colony and has since been reported also for rhesus monkeys under semi-wild conditions (ALTMANN, 1962) and for savanna baboons in the wild (DEVORE, 1962). In contrast, hamadryas baboons in the wild almost never resort to the protected threat, and then only in a much reduced form:

During a quarrel between two adult females, one female places herself squarely between her rival and her leader, without, however, presenting to him. She screeches, at the same time glancing alternately at her leader and her rival. The male, however, does not react to her behavior.

The distribution of the protected threat pattern illustrates the mistake in the view that a field study will reveal all the 'relevant' behavior of a species. In fact, it only reveals one modification of special biological interest.

e. Relations Between Adults of Different Units; Aggression Between Leaders

There are few interactions between adult unit members and non-members (Table VIII, p. 80). The sociograms show that about 45% of the adult individual's time is spent in interaction with members of his own unit, while only about 3% is spent relating to outsiders.

For troop life, the most important contacts between the units are the short interactions between their leaders. Grooming was never observed among leaders of mature units. Whereas adult males having no females often groom each other, the unit leaders do not approach each other at grooming distance, but ordinarily keep at least 1.5 meters apart. This spatial separation appears to be related to the leaders' strong tendency to keep their units from intermingling. The ordinary distance is only understepped during two types of interaction. The first type serves to coordinate troop movement and is cooperative rather than aggressive. It will be dealt with in the chapter on troop organization. The other type of contact observed between leaders is aggression.

Fighting technique consists of each opponent aiming bites at the shoulder or neck of the other (Fig. 19). Among hundreds of such scenes we have only seen a male actually take hold of another's coat on two occasions. The analysis of films shows that the animals fence rapidly with open jaws without really touching each other and that the heads are often held back. During a fight each opponent also hits out at the face of the other with his hand, usually missing here as well. The biting and hitting ritual goes on with tremendous speed for a few seconds, silently, the opponents facing each other. Then, one of them turns to flee. At this moment the other often snaps out at him, producing an occasional scratch on the anal region. The vigorous mutual chasing, interrupted by some more fencing, usually lasts no longer than 10 seconds. Most fights come to an end when

Fig. 19. Two adult males fighting.

one of the opponents flees. Nevertheless, a fight can be abruptly stopped, even in the midst of dueling, if one of the opponents suddenly gives the other absolute advantage (Fig. 20 b):

During a biting match, one of the animals suddenly turns his head, exposing the side of his neck to his opponent. But instead of taking advantage of this, his rival immediately stops snapping at him.

From a film strip: During a pause in a fight between two unit leaders, one of the opponents starts to approach the other with rapid pumping lip movements (inflating his closed mouth between chewing movements). In response to this threat, the other backs off for a few paces and turns his head aside, thus exposing his throat to the jaws of the oncomer. Instead of biting down, however, the opponent 'yawns' at the other's neck and with a few decreasing chewing movements, backs off.

LORENZ (1952) has described a similar, but more pronounced, gesture which releases a similar inhibition in dogs and wolves. By directly offering his throat to the bite of his opponent, a losing wolf can prevent this bite and stop the fighting; the superior fighter remains motionless, his teeth poised at the proffered neck. For the hamadryas, as well as for the wolf, a direct bite in this region is likely to be fatal. Apparently, similar weapons and fighting techniques in the two species have convergently led to a similar inhibition, released by a similar gesture. This may partly explain why we never saw an animal gravely wounded on the neck, or killed. Although most bites are aimed at the neck, it appears that more injuries are inflicted on the hands and forearms. On about ten occasions we came across males who were unable to walk on all fours, while lame females were encountered three times.

Duels between hamadryas males occur only at familiar resting places and easily spread out to include other males. Only when a battle lasts for several minutes are sides distinctly formed (p. 104). Usually, however, it looks as though chasing males run chaotically through the troop, here and there inciting other males they have accidentally come too close to chase them. For the rest, neighboring males just watch, roaring a deep 'Oohu – Oohu' (the second syllable of which is inhaled).

There is no clear relation between the size of an opponent and his tendency to flee. We have seen a subadult male chase away three adult males in a single attack. Nor is the number of females indicative of a male's success in an encounter. Smoke, for example, who possessed three females, ran away from a male who possessed only one.

Fights between two males often start when a male attacks another who was actually not threatening him, but a third animal.

An adult male A directs his attack at a juvenile, who immediately backs away, but is nonetheless grasped by the neck and shaken by his attacker. A neighboring leader B, yawning with his head protruded, now attacks the two. By mere chance, it seems, this attack goes also in the direction of a third male C who immediately attacks B. A heavy duel between B and C follows, during which both males roll over the side of the cliff.

Apart from three such cases, all of the thirteen fights that we observed from their very beginning, revolved around the possession of a female. A typical example of this is the following:

From a film strip: During a general melee which we cause by placing food at the sleeping rock, an adult anoestrous female carrying her infant on her back is seen running away from a male A who is chasing her. When his jaws are still about a meter away from her, male D attacks him from the side (Fig. 20a). The female escapes as the two of them begin to fight. D succeeds in biting A who falls to the ground. The female now runs past the beaten A to D who, it appears, is her leader. As she crouches in front of him, D stands over her and takes hold of her hair. A then backs away and a subadult female in oestrus comes over and presents to him.

In the meantime, two other adult females and one subadult gather about D. A few minutes later, A again attacks D's female who is still carrying her infant and again a fight between the two males takes place. This is repeated a third time. Then, A approaches D with open jaws. D looks away exposing his neck (Fig. 20b); A brings his teeth about 15 cm to D's neck, then walks away.

Evidently here the attacker A was interested in a female belonging to another unit leader D. But instead of attacking the owner D he attacked the desired female; possibly in order to give her a bite on the back of the neck. Only when the owner interceded did a fight between the two rivals take place. It is to be remembered that if a female in oestrus copulates with a male outsider, the leader does not attack the outsider, but the female. If, however, an outsider tries to take over a female by force, then the leader will direct his attack at the usurper.

None of the fights we observed was brought about because of females in oestrus. This further supports the notion that in hamadryas males, the permanent possession of a female is not primarily motivated sexually. Savanna baboons, in contrast, do not fight over anoestrous females. They only harass each other over the temporary sexual association with a female in oestrus.

'Foreign relations' of females, aside from their copulatory activity, are less conspicuous. Here and there they will fleetingly present to strange adult males, and in two instances we have observed a female temporarily attach herself to another unit.

A female in oestrus, after intensively presenting to a strange unit leader, follows his unit briefly.

A strange female sits down in Smoke's unit on the sleeping rock. After a while she approaches Smoke and pulls his hair. But he does not react to this. Later, she is groomed by one of Smoke's females until her own leader comes to her and threatening her, induces her to return.

As a rule, however, there is no contact between adult females and strange adult males. Between females of different units short hesitant relationships arise motivated by an interest in small infants. Adult and juvenile females from neighboring units will approach the mother of a small child, gaze or sniff at it for a few seconds and then withdraw.

Two females from different units, each bearing a small black infant, approach each other. Each of the mothers places her nose close to the other's child and then carries her own infant back to her unit.

In savanna baboons, the most frequent type of temporary subgroups is a cluster of females assembled around a mother and her

20 a

20 b

Fig. 20. (a) Male D (center) attacks male A, who is chasing D's female (on the right). (b) A approaches D with open jaws, but D interrupts the fight by exposing his neck. The female drops in line behind D.

young infant. The tendency is obvious among hamadryas females, but the intolerance of leaders makes actual clustering of mothers impossible. Only rarely does a female groom a strange female or a juvenile female who may have gotten close to her while inspecting her infant. No female was ever concerned with a strange infant after it had outgrown the black-haired stage.

Finally, there are sporadic exchanges of threats between adult animals of different units. On the sleeping rock, preferred cornices and shelves are usually left to older animals without further ado. Fights were never seen to start because of sleeping space. Most of our identified one-male units always took the same station on the sleeping rock night after night. In this way each leader usually had the same neighbors. Individual sleeping places, however, might change frequently as a result of competition among the neighbors; but here, too, these changes occurred in an area whose diameter was usually no more than three meters.

f. The Followers

The relation of a young male follower to a specific one-male unit takes three forms, related to the follower's age. The first form or phase is typical of juvenile males, the second of subadults and the third of young adults.

In the first form, infant and juvenile males are often found near one-male units with females in oestrus and tend to follow the unit at such times:

At the end of a 'copulation' series with a 1½-year-old male, a two year-old female is caught by her young leader. Attacked by him, she runs up a tree, screaming, followed by the 1½-year-old male. Later, when the leader breaks camp with the troop, the female follows him in response to his glance. The 1½-year-old male, in turn, follows her.

A three-year-old female belonging to the young adult male 'Guy' is in the second half of her oestrus. As the unit enters the sleeping rock, she is followed by two young males, a subadult and a 1½-year-old. Guy repeatedly copulates with her; in between threatening the younger of the two males, yawning, lifting his brows and hitting out at him. The young male shifts back and forth uneasily and scratches himself, but does not leave. Later, he approaches the female and looks into her face.

Thus, it appears as though it is the females in oestrus who introduce the young followers into the units. We have already seen that the latter do not lack in opportunity to copulate.

During the second phase, copulation between the now subadult follower and the females of the unit becomes infrequent. Nevertheless, followers stay with the unit for longer periods during this phase, even when the troop is on the move (Fig. 21). On the sleeping rock itself, the subadult followers are often extensively groomed by females of the unit to which they have attached themselves. Such grooming between females and followers is tolerated by some leaders; others respond with a slight threat:

Two followers are being groomed by Smoke's females. After being gazed at for some time by Smoke, the younger follower becomes restless and withdraws, although he has not been threatened. His grooming partner utters a contact-grunt and turns her grooming attentions to Smoke. Later Smoke is seen pumping cheeks at the other couple, causing the subadult follower to turn his back abruptly on the female and thus break off her grooming.

When a subadult follower is actually threatened by the leader, he may behave as would a female in the same situation, i.e., he runs to the leader.

A unit leader rises and stares at his subadult follower who is groomed by one of the leader's females. At this, the follower screams and runs over to the leader, who mounts him. Shortly thereafter, he follows two adult females of the unit to a spot away from the leader.

Fig. 21. A young subadult follower (rear) travelling with his one-male unit.

Fig. 22. The subadult male on the right attached himself to the unit of the young leader (Guy) in the center, but was threatened when he tried to approach the females. He finally left the unit for good. The leader's 'yawn' is related to the follower's presence.

That no followers were counted in the initial units of the broad sample suggests that young leaders are especially intolerant of followers (Fig. 22). In contrast, we will find the old leader Rosso more tolerant than the prime leaders described here.

In addition to these slightly aggressive encounters with the leader, the relation between a leader and his subadult follower may show the first signs of that 'notifying' behavior which plays a part in the coordination of traveling; i.e., they look at each other or present to each other before they change their relative spatial position (p. 128). Only a few subadult males have a notifying relationship with their leader. Sepp, a subadult follower of old, influential, but very tolerant Rosso, had such a relationship.

During the midday rest at a watering place, Sepp is sitting about a meter away from Rosso. Rosso looks over at Sepp who turns his face to the leader. After some thirty seconds, Rosso rises and another glance, together with contact-grunts, is exchanged. Sepp rises too and slightly bares his teeth. Rosso gazes downstream but does not depart. Sepp turns his head and also gazes downstream. Both animals then sit down again.

A little later Sepp rises and begins to walk towards the water. But after having gone about two meters, he turns and looks at Rosso, then comes back and presents to him with lifted tail. Rosso briefly touches Sepp's scrotum with his hand and Sepp sits down next to him. After a minute has elapsed, Sepp actually goes down to the water, where he takes a leisurely drink. Three minutes later, he comes back and sits down close to Rosso.

That Sepp bares his teeth at Rosso's glance and that his tail is raised as he presents indicates that his interactions with Rosso still evoke some flight tendency in him; a component which is not obvious in a full-fledged notifying relationship.

Among young adult followers this notifying relationship to the unit leader is fully developed. In this, the third phase, the follower's grooming relationships to the unit's females have stopped entirely. Thus, the adult follower 'Wave' engaged in grooming only with his leader Rosso. His sole interaction with the females was a single intention movement of presentation by one of them. While subadult males generally run along anywhere in the unit's neighborhood during a march, adult male followers usually bring up the unit's rear. If the follower moves off to the side, the females fan out, in accordance with their centripetal tendency, between him and the leader. Even the direction a unit takes is somewhat influenced by the adult follower, and he and the leader are regularly observed notifying each other. Thus the adult follower may be considered a second, subordinate unit leader, with one important reservation: He has no access to the unit's females. They belong to the leader and avoid coming too close to the adult follower:

On route, Rosso, closely followed by his females, directs his steps towards his adult follower, Wave. Rosso notifies Wave by presenting and then steps aside in the characteristic rapid turn. In this way, the inattentive females are suddenly left facing Wave. This startles the foremost of them so that she jumps back uttering a short 'ha' (vocalization correlated with the sudden appearance of an unfamiliar object).

While subadult followers frequently leave the unit during travel, the adult follower 'Wave' followed his unit at all times and actively sought it after he had lost it (p. 148). Possibly, this relationship between leader and adult follower later develops into a two-male team.

Thus, we have seen that the role of unit follower changes from the occasional sexually motivated visits of a juvenile male to a unit's females into a follower-female grooming relationship and finally develops into a close relationship between the adult follower and his leader. Although these phases are related to the follower's age, there is nothing to suggest that each young male goes through all of the three phases, nor that all of the phases occur with the same unit, although this is conceivable at least for phases two and three. Also it is not known whether the followers are originally strangers to the units or sons of their members; though some of the leaders were too young to be the fathers of their followers.

g. *Infants and Juveniles in the One-Male Unit*

The infants and juveniles of the identified units spent an estimated 60% of the observation minutes in social interactions. About one-third of this social time was spent with baboons outside the youngsters' maternal unit, whereas adults spent less than one-tenth of their social time with animals outside the unit. Older infants and juveniles freely leave their mothers' unit. According to the criteria given on page 29, they therefore are not genuine members of the one-male unit. We have already seen that young males are found in strange units having females in oestrus and that juvenile females in oestrus often wander through large portions of the troop.

As long as an infant still has its black coat, a mother will be intensely occupied with him. In the first weeks of life he is kept on her stomach the whole time. If she happens to be walking, the young animal clings tightly to the hair on her sides (Fig. 23). Many mothers of black infants have a distinct bald triangle on their flanks, as a mark of the infant's grasp. Even a dead black infant was carried on the belly of his mother. On only one occasion have we observed a mother laying her ten-day old infant on the ground during a march.

It began to scream immediately. A two-year-old male ran up to it and presented his lowered hindquarters to the infant, inviting him with this typical gesture to climb upon his back. This, of course, met with no success, since the infant was not yet able to walk. After this, a subadult female offered her belly to him, but also without success. The infant then crept some 30 cm back to its mother and clambered onto her belly.

As soon as they are a few weeks old, black infants are seen crawling to neighboring peers or tottering sometimes as much as a meter away from their mothers, watched all the while by the adults

Fig. 23. Two mothers with black infants climb to their sleeping ledge.

sitting about. Still later, when the black infants are able to walk, they will respond to an invitation of the turned dropped rear by jumping on the proffered back. In this way they are often brought back to their units by their mothers. A sex difference is already apparent at this age in the data of the broad sample: Black males leave their mothers more often than do black females (p. 91). Mothers do little to control the excursions of older black infants; they even tolerate them to climb on small trees and on the cliff to the very limits of their capacities, and to stray up to about 3 meters away from the mother.

All age classes approach black infants with interest. Juvenile and adult females will go up to them and sniff at them. (Hamadryas baboons will sniff at objects held by other individuals, or at any objects that attract them but which they are afraid to touch [Kummer, 1957]). Mothers almost always permit their infants to be sniffed at, but rarely allow them to be touched.

Slender, an adult female, is seen approaching a nursing female of a neighboring unit. First she sniffs at the infant and then reaches out to touch him. The mother immediately hits her at this; Slender bares her teeth and withdraws.

Other mothers are more tolerant:

A mother offers her infant's rear to her female neighbor and joins her in examining it.

An adult female in oestrus touches a strange black infant and lifts him to her mouth. When he begins to squeal, its mother immediately takes hold of him and draws him against her belly. Shortly thereafter, the infant crawls over to the other female of its own accord. The female in oestrus then presses the child to the ground and again the infant begins to squeal. Once again, his mother takes him back without making the smallest threatening gesture. Later the infant begins to wander towards an adult male but is grabbed by a 1½-year-old female. The mother rises to follow her male; directing an intended neck-bite at the 1½-year-old female, she takes her infant and carries it away against her belly.

Female care of an infant other than her own was seen only once:

Two females are sitting next to each other at a watering place, when suddenly the black infant belonging to one of them slips into a crevice. The other pulls the child out and shoves it against the breast of its mother.

On route, black infants are sometimes carried on the backs, and, more rarely, against the bellies of adult males, while their mothers walk along next to them.

On the move, an adult male turns and looks at me, grabs the black infant at his side and carries him away against his belly. The child's mother walks quietly beside him.

As long as the infants' coats are black, they may crawl over the knees and onto the heads of adult males, who, in turn, may hug them but never threaten or attack them (Fig. 24). When the infant has lost its black coat, between the ages of 4 and 6 months, it loses its preferred status. Above all, there is a change in the attitude of the mothers. Although brown infants are still frequently groomed by their mothers and are always carried by their mothers during flights and fights, they are rarely ever brought back if they happen to wander off to another unit. In the mother-child relationship, the initiative now passes to the child; the mother is usually more passive (Fig. 25).

A squealing one-year old tries to overcome a difficult spot on the sleeping rock. Its mother merely stands and watches as he tries to climb over the obstacle. After a moment he finally meets with success and quickly runs over to his mother. Still humming softly, as forsaken infants do, he climbs onto her back.

One-year-olds who try to suck their mothers' nipples are sometimes bitten by their mothers. Nevertheless, even a two-year-old has been observed holding a nipple in his mouth while his mother groomed him. When the troop leaves a resting place, the one-year-olds still return

Fig. 24. An adult male without a unit hugs one of a group of black infants which assembled to play around him.

from their play groups and join their mothers. The mothers may look out for them before they follow their leader:

> Break of camp. Smoke is seen rising and starting on the march with his females. After a few moments, Slender, his female, stops and looks back. Forty seconds later, Smoke comes to her and looks back as well. Now Slender's one-year-old daughter comes out of the bushes, squealing. At this, Smoke immediately continues on his way with the other females while Slender and her daughter catch up.

The youngsters' relationship to adult males also changes with the change in the infant's coloring. Aside from adopted children (p. 114), adult males no longer carry brown infants, and the male's tolerance for them is much less than for the black infants. A unit leader even bit a one-year-old when he tried for a second time to squeeze between the leader and a female. Young animals at this age often respond to glances from adult males with squealing, but will also run to sit between their knees when attacked by other animals. Adult males are

Fig. 25. After vainly trying to descend over a difficult passage, a one-year-old has started to scream. His mother finally returns and offers him her back as an additional step.

potential aggressors as well as potential protectors to the young brown juveniles:

A young animal, barely a year old, squeals and crawls away from an adult male. The leader of the infant's unit now approaches the first male and threatens him, raising his brows and stretching his head forward. The infant, turning to his leader, screams and crouches to the ground. Thereupon, the leader reaches into the bushes, pulls the infant out and returns to his unit with it. The infant then jumps up on one of the females and the unit breaks camp.

Several juvenile animals and infants are threatening two vervets (*Cercopithecus aethiops*) who are two meters away. An adult male then comes over to them and as he is making a mock attack at the intruders, one of the juveniles jumps on his back.

Up to the age of about 18 months, the juvenile regularly returns to his maternal unit, even though the juvenile male already spends much of his time in the play groups that gather in and around the resting troop (Fig. 26). Occasionally, he even will spend the night in a group of his male peers, away from his mother. The female of 18 months may already become the first consort of a young leader.

Fig. 26. A group of wrestling juveniles.

Whether she can permanently remain in her mother's unit is not known; but she is always found in a one-male unit and leaves it less and less frequently as she grows older.

h. Kidnapping and Adoption

Subadult and young adult males who do not lead a one-male unit tend to take black infants away from their mothers for periods up to 30 minutes (Fig. 27). In the 'unit of mothers' (p. 78), led by the male 'Sweep', there were five infants, all in the black color phase. The following 'kidnapping' scenes took place in this unit:

One of Sweep's females stands facing a bush and screaming. Sweep stands up and threatens in the direction of the bush. As he rushes toward it, a subadult male dashes out of the bush, carrying a black infant on his back. Chased by Sweep, het kidnapper takes flight over the rocks to another part of the troop where fighingt breaks out. After a few seconds, Sweep's female is seen inviting the black infant to jump on her back. Immediately, Sweep leads his unit back to its previous place.

Ten minutes later another mother in Sweep's unit goes to the edge of the cliff and looks down at the shelf below. Unseen by Sweep, a subadult male is there,

Fig. 27. A mother looks back at an older subadult male who has kidnapped her infant.

carrying a black infant on his back. The female, glancing first at Sweep, then at the subadult male, walks along the edge of the cliff, keeping pace with the subadult. Time after time, the infant tries to climb up the cliff but is always pulled back again by the kidnapper who then stands over him with open jaws. The infant begins to hum and is answered by his mother's contact grunt. As the subadult male carries it further, the infant begins to squeal. The mother paces back and forth, and glancing back at Sweep she utters a staccato-cough, which is answered by a contact grunt from Sweep. Upon receiving Sweep's contact grunt, she begins to climb down towards the infant. Halfway down she returns, stands bipedally on the edge of the rock and looks towards Sweep, then returns to him. The subadult male keeps the infant for another ten minutes, inviting it to climb on its back by lowering his hindquarters, hugging it, carrying it back and forth and trying to get it to join in a play fight. The mother proceeds halfway down the slope twice again, but only goes all the way down when this is finally possible without losing sight of Sweep. Without threatening the subadult male, she follows him to her departing unit where the infant changes over to her.

A subadult male simultaneously handles two such infants. One is clinging on his back; the other to his belly. The mother of one of the infants follows him, sniffing at her child.

Similar scenes can often be seen during the birth season. The kidnappers are almost always subadult males, rarely young adults. Both male and female black infants are kidnapped.

The tendency of subadult males to appropriate and handle black infants is the earliest of several forms of 'maternal' behavior in young hamadryas males. A second form, adoption, was found when we released infants and young juveniles into foreign troops. All these animals were immediately caught and mothered by young adult males (p. 114). When a hamadryas mother dies leaving an infant, it is probably not the unit leader or another female who takes care of such an orphan, but one of the young males. Observations on an identified male of about 2 years suggest still another function of adoptions. This male still occasionally fled to his mother when frightened, yet at night he slept huddling against the chest of a neighboring subadult male. Giving up the relation with the mother may thus be less difficult for a hamadryas infant, who will easily find a male adoptor, than for a young anubis baboon.

The third form of the young male's 'maternal' behavior is at work in the foundation of the initial unit, which we shall now describe.

3. EARLY AND LATE STAGES OF THE ONE-MALE UNIT

The preceding sections have described one-male units at the prime of their leader's life. The leader's behavior, however, changes with his age, and the composition and activity of his unit are so strongly centered around him that they go through a corresponding sequence of changes. The following descriptions trace these changes in chronological order, thus reconstructing the probable life cycle of the average one-male unit.

a. The Initial Unit

In the broad sample (Fig. 14) many units were found that consisted only of the leader and one infantile or juvenile female. All leaders of such units were young adults. This, and the observed development of an equal unit in the zoo colony, suggests that many young males begin their careers as leaders with a single infantile female. This hypothesis has led us to designate such units as initial units; it is supported by our close studies at the White Rock.

Four initial units were usually found among the White Rock troop. All of their leaders were young adults with dark faces and mantles still partially brown. Their females were 1 to 1½ years old, which

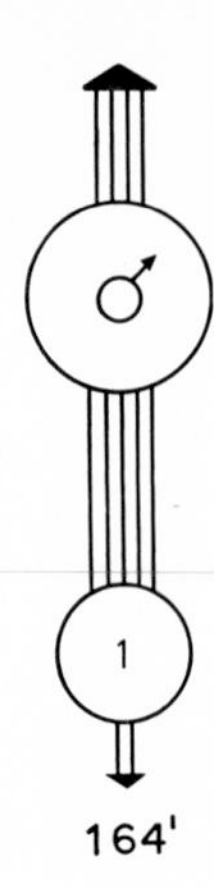

Fig. 28. Sociogram of Naso's initial unit. The male-female distance is 41 cm. For key, see Fig. 15.

means that they were at least 2 to 3 years from child bearing age. One such leader was 'Naso' (Fig. 28), who was observed, together with his one-year-old female, 'Pluck', for a period of 58 days. Their behavior was frequently compared with that of the other three initial units. Like the older males, the leaders of the initial units herd their females and are groomed by them (Fig. 29). The young couples even show an especially marked cohesiveness. Naso, for example, would always look back at Pluck when preceding her. A look from him was enough to evoke a staccato cough from her and bring her to his side. However, he never gave her a bite on the neck. At most, he would draw her to him with his hand, as mothers draw in their small infants. In the other initial units only one neck bite was ever observed. To invite grooming, Naso would present his flank to Pluck in the manner of older leaders. Pluck, for her part, would groom him frequently, though tiring quickly.

The first characteristic that distinguishes these units from the units of older males is the absence of sexual behavior. The female, at this age, is not yet sexually receptive. More striking, however, is the appearance of forms of behavior peculiar to mother-child relationships. For example, Pluck would run into Naso's arms or cling to him during a fight between two neighbors (Fig. 30). Another male was seen taking his female on his back because she had screamed as some baboons crowded past her on a narrow ledge. At difficult passages on the sleeping rock, which the little females could not

Fig. 29. The juvenile female of an initial unit grooms her leader.

negotiate, they would sit down after several unsuccessful attempts and begin to hum softly in the manner of lost infants. At this, their males would often return and pick them up on their backs to carry them over the difficult spots (Fig. 31). In mature one-male units, neither adult nor juvenile females are carried by the leader. Nevertheless, adult females occasionally regress to infantile behavior in that they cling to their leader when they are threatened by other baboons.

The leader of the initial unit thus handles his one-year-old female with patterns of maternal behavior which she would otherwise receive from her mother. Maternal behavior in males is observed sporadically in the other baboons and in macaques (e.g. ITANI 1963). In the hamadryas male, however, it appears to assume a major function. During the years which immediately precede the formation of a unit, the maternal behavior of the hamadryas male develops systematically in three ontogenetical steps marked by an increasing specialization towards a permanent bond with a young female: The subadult male kidnaps infants of both sexes for a few minutes only; the young adult male adopts juveniles of both sexes, but for

Fig. 30. During a fight among neighbors, a juvenile female cuddles to her leader's back.

a b

Fig. 31. A leader offers his back to his one-year-old female (a) and carries her over a difficult passage (b), which she vainly has tried to cross.

a more prolonged relationship; finally, the young leader adopts only female juveniles and thereby founds his one-male unit. Before and after these three periods, maternal behavior appears as sporadically as it does in the related species. This ontogeny suggests that male maternal behavior toward females may be the ontogenetical and evolutionary raw material of the hamadryas one-male unit. It may establish the stable bond between male and female which sexual motivation alone fails to create in baboons.

Initial units are immature in still another respect. Both partners still carry on types of social relationships which are characteristic of the period before the founding of the unit. Thus, as seen in the sociogram, Naso spent about 10 times as much time with animals outside his unit as did the leaders of other units. During this time, Naso would engage in prolonged mutual grooming with 'Prince', the leader of another initial unit (Fig. 32). This relationship with Prince was presumably carried over from the time when both males lived together in a cluster of subadult males. Furthermore, both of them continued to sleep on the same ledge with their child-females in a gathering of juvenile and subadult males. The following illustrations

Fig. 32. Grooming between the members of two initial units (Naso and Prince). Frequent cross-unit interactions are typical of initial units.

show Naso's close relationship to a group of juvenile and young adult males, and his response to Pluck's climbing difficulties.

17.43 h Naso and Pluck have just come down onto the upper edge of the sleeping cliff. Naso sits beneath a difficult overhang while his female is still above. Naso looks at her and she utters a staccato-cough. As Naso climbs further down, Pluck loses sight of him. Staccato-coughing, she now tries to climb over the overhang. She does not succeed.
17.47 h She runs along the overhang until she comes to stand two meters above Naso. Staccato-coughing and beating her tail from side to side, she looks down at him. He glances at her briefly.
17.50 h Pluck stands up and begins pacing back and forth along the edge. She scratches herself and sits down again.
17.52 h Again, she stands up. Naso responds by looking up at her. She staccato-coughs and patters in place. Then she sits down once again.
17.55 h Naso climbs over to the sleeping ledge No. 8 and is groomed by one of the two-year-old males sitting there.
18.00 h Naso and Prince are engaged in grooming one another.
18.10 h Pluck, still sitting on the same spot above the wall, is dozing.
18.18 h A strange adult male sits down within 60 cm of Pluck, whereupon Naso looks up and Pluck begins to staccato-cough.
18.20 h While Pluck watches, Naso climbs down to the foot of the cliff and circles the sleeping rock until he reaches the upper edge of the overhang. Even before Pluck is able to see him, she begins to go towards him along the edge. As soon as he appears, she staccato-coughs and runs towards him.
18.32 h Screaming, she follows Naso down another path in the cliff. Finally they settle down on sleeping ledge No. 14 where she immediately begins to groom him.

Thus Naso took part in relationships with his neighbors at the first sleeping ledge and gave these relationships up in order to lead Pluck to a place which she could manage.

The following evening Naso tries again to reach ledge No. 8, but this time from the bottom of the rock. Again Pluck is unable to follow and Naso tries, unsuccessfully, to invite her to climb upon his back three times. Finally he leads her, after all, to ledge No. 14.

The females' relationships to outsiders were as immature as their relations to their leaders. Pluck's main outside relationships consisted of playing with females of her age. This relationship came into being whenever Naso was busy grooming with Prince. Prince's 1½-year-old female even went so far in overstepping unit boundaries as to groom Naso. Thus, the social isolation typical of the mature one-male unit has neither yet fully developed in the male nor the female of the initial unit.

Some observations give an indication of the way in which an initial unit is formed. The leader's behavior in the process is similar to the kidnapping practice and almost identical to the practice of adoption (Fig. 33).

During the break of camp, the one-year old female of an initial unit is seen repeatedly fleeing from her male some 10 to 20 meters into the brush. Each time the male chases after her and brings her back out. She then follows him to his place, screaming all the while.

Strong motivation to flight was still evident in Pluck's behavior towards Naso, even though it never resulted in a full-speed escape. Again and again she would begin to slide away from him slowly, while still sitting, her eyes fixed on him the whole time. But then, within about a ½ meter, she would begin to staccato-cough under his gaze or run back to him screaming. From time to time, she would indiscriminately try to approach one-male units that passed; the actual approach was always prevented by Naso's staring at her.

Such observations lead us to suspect that young females at first tend to flee from their leaders and that they develop the following response only under many repeated attacks from the leader. The juvenile female apparently must be conditioned to follow a leader. Later, the adult female easily transfers the following response to another male (p. 109). The retrieving attacks in the initial unit are remarkable in that the young females are hardly ever bitten on the

Fig. 33. The leader of an initial unit attempts to grab his escaping female.

neck but caught with the hands, although the male may become so agitated that he barks in alarm. This is apparently so because there is an inhibition operating in the male which prohibits the biting of small animals (KUMMER, 1957). At times, this inhibition breaks down. The probable effects of this were seen in three of the four females among the *initial units* of White Rock. All three bore deep perpendicular cuts running from the ear or from the corner of the mouth to the crown of the head. Since we have never seen a similar injury on an adult female, we may conclude that in an occasional neck bite the small heads of these infant consorts were partly encompassed by the huge jaws of the males, and that in slipping off, the rear edge of an upper canine left its trace (Fig. 34).

How a young adult male first removes a juvenile female from the unit in which she was born is not known. Since we never saw a troop member interfering with a young male's chasing a juvenile female, there is nothing to suggest that the female's original leader, who will usually be her father, opposes the removal of his daughter. As we have seen earlier, mature leaders having adult females rarely interact with juvenile females of their unit and do not prevent them from wander-

Fig. 34. A juvenile female with a typical scar on the head presents to her leader.

ing through the troop; also, the mature leaders show no tendency to kidnap or adopt infants or juveniles. As Dr. CLAIRE RUSSELL suggested in a comment on this section, the mature leaders' indifference to their juvenile daughters, together with the young males' tendency to appropriate exactly such juveniles for their initial units, may considerably reduce inbreeding between fathers and daughters, which might otherwise be very frequent in the hamadryas society.

For about one year after the initial unit is established, the young female has no monthly sexual swelling, and copulation is therefore impossible. Already at the age of two, however, some hamadryas females start to develop occasional swellings and, if they live in initial units, to copulate with their leader, whereas female savanna baboons have their first swellings only at 3.5 to 4 years of age (HALL and DEVORE, 1965). Nevertheless, no hamadryas females under the age of 4 years were observed with infants of their own. It also seems physically impossible that a small two-year-old could give birth. A reproduction physiologist might investigate the physiology of the premature swellings in juvenile hamadryas females and the hypothesis that their early sexual receptivity may be an effect of their association with an adult male who has no other females.

b. The Transitional Stage

Guy was a young adult male whose coat was still mostly brown and whose canines were long and intact. He and his three-year-old female 'Suzie' may have formed a typical initial unit somewhat more than a year before the start of the observation period. Now, Guy had a second female, 'Kit', who was only two (Fig. 35). During the 40-day observation period, Suzie was twice in oestrus, but had never yet suckled, as far as I could tell from the shape of her nipples. During the first week of the observation period Kit, the two-year-old female, underwent a period of swelling synchronously with Suzie, but had no swelling in the second month. Her youth and the fact that Guy had copulated exclusively with Suzie the preceding month excludes pregnancy.

Relations between Guy and the older of the two females had already entered into a more mature phase. Mother-child behavior was no longer apparent and sexual behavior was fully developed. Guy and Suzie had reduced their relations with strangers to the 2% of the

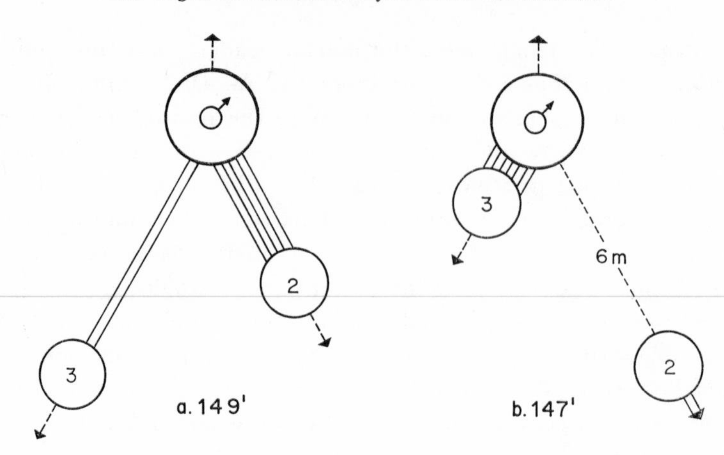

Fig. 35. Sociogram of Guy's unit, (a) when the older female Suzie was in anoestrus and (b) when she was in oestrus. Male-female distances are 97 and 41 cm in (a), 9 and 600 cm in (b). For key, see Fig. 15.

observation minutes which is typical of a unit in its prime. Social behavior was focused to an extreme degree on internal matters of the unit. In fact, Guy controlled his females with an exaggerated intensity. Almost every evening he would lead his two females on a 'following-procession' over a routine detour around the sleeping rock to the unit's sleeping ledge, and although little distance was covered, the march would usually take more than 20 minutes. During such processions Guy would look around at every step or so and stare at his hurrying females. Every few meters he would stop while his females would quickly scamper to catch up with him. Then there would be intensive grooming either by him or his females. Sometimes he would press Suzie to the ground as she approached and bend over her with chewing movements. This gesture has only otherwise been observed during fights over the possession of a female (p. 102). Sometimes all the animals on Guy's side of the cliff would already be sleeping on their ledges while he still led his females about in great agitation. We were unable to discover any external causes for this exaggerated tendency to lead and herd. Following processions were not observed in initial units nor in units belonging to old males; though they were seen in a less intensive form among leaders in early prime.

In contrast to Suzie, Kit, the younger female, behaved like a female of an initial unit and was treated accordingly by her leader.

Guy would invite her to be carried on his back by lowering his hindquarters and would often embrace her while he sat. When grooming her, he would sometimes lift her by one leg as mothers often did their children. Another unit of the same composition changed its sleeping place when the smaller of the two females was unable to climb over a difficult spot. Kit, just as Pluck did in Naso's initial unit, responded to her leader's gaze by staccato-coughing, coming up and presenting to him or grooming him, whereas Suzie, the elder of the two, did not always comply, even when clearly threatened.

Both of Suzie's periods in oestrus caused basic changes in the unit sociogram. During her anoestrus (Fig. 35 a), Guy would watch over both females. He was able to keep Suzie within 1.3 meters by threatening her often, but was seldom groomed by her. At the same time he could keep Kit much closer to him simply by casting glances at her. She would groom him almost after every shift in location, but like Pluck would tire after about two minutes.

During Suzie's oestrus (Fig. 35 b), however, Guy watched only Suzie, who now more often approached to groom him. During this time, Guy threatened any male who seated himself near her. Such threats by eager young leaders probably keep potential followers away from young units. During Suzie's oestrus, Kit was able to leave the unit altogether without Guy's interference. She would play with peers at quite a distance and several times sought out an adult and a three-year-old female who perhaps belonged to the unit of her infancy. Now and again she would return to Guy and present to him.

Guy's unit may be characterized as a young unit of a young leader who, having broken off relations with his peers, has now turned his attention to leading his females with great intensity. The older of his two females, although not completely grown, has already assumed the normal female role and is under constant supervision. The relationship of the younger female, including her excessive response to threats, still bears characteristics of the initial unit. This strong response forces the young female to 'take the place' of the older one close to the male during the older's anoestrus, while at other times she may still enjoy the freedom of a juvenile.

The unit belonging to Fork, also a young male, was studied over a period of six months (Fig. 36), but showed no essentially new characteristics and will therefore not be treated in detail. Fork

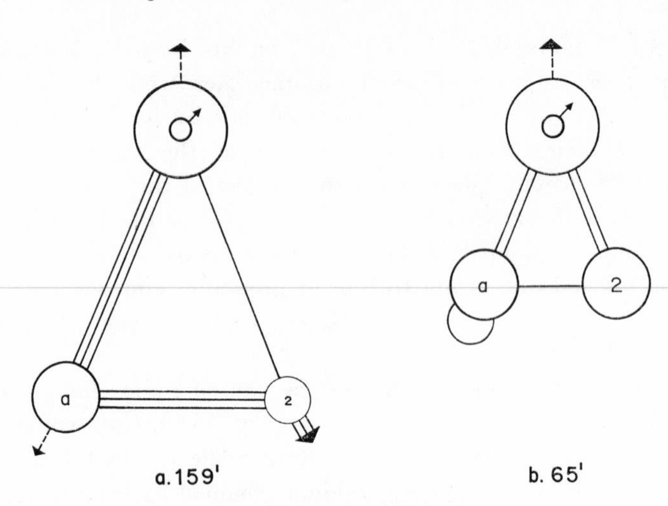

Fig. 36. Sociogram of Fork's unit, (a) before and (b) after the adult female had given birth. Male-female distances are 93 and 94 cm in (a), 53 and 52 cm in (b). For key, see Fig. 15.

appeared to be reluctant to assert his leadership and followed his one adult female instead of threatening her. He sat down on an average of 48 centimeters away from his female (n = 10), whereas, on the average, she settled at nearly twice this distance from him (84 centimeters, n = 13). Her grooming was limited to the times when she returned to him from a distance and even then continued only for three to four minutes. Fork made some tentative movements to undertake following processions:

Fork's female leads the way to the sleeping ledge. After passing hesitantly in front of an adult male, looking into his face all the while, Fork runs to his female and both go to their sleeping ledge where she grooms him rapidly for three minutes. As she stops grooming, Fork moves some 80 centimeters away without taking his eyes from her. Whereupon, still seated, she slides after him and answers his gaze for five seconds. As she lifts her hand to his head as though to groom him, he again slides away from her in a sitting position, and she again slides after him. This time he remains still and she begins to groom him.

Fork's female already had a female child whose age at the beginning of our observation period was estimated to be 1½ years. It followed her persistently and although it was obviously her daughter, it behaved towards Fork more and more as though it were one of his females. This was especially the case when her mother began to pay him less attention following the birth of a new infant. The age make-up of the unit shows that Fork could not have acquired his female in her infancy, but rather took her over, together with her daughter, from another leader. Thus the unit would not have gone through the stage of initial unit.

The new infant was born towards the end of the observation period. Copulation had evidently not occurred during the entire six months. Despite this, the unit had remained together unchanged.

c. *Units of Leaders in Late Prime and Old Age*

The next stage in the development of the unit is characterized by a rapid increase in the number of adult females, most of which are probably taken over from old leaders. The leader is now in his early prime and his unit has reached its zenith as we have described it in earlier sections. With a description of Rosso's unit (Fig. 37), we turn our attention to a unit whose leader is in his late prime. We shall then go on to the units of Pater and Silver as examples of units whose leaders are aged.

Rosso's face was, as is typical for older males, bright red and his silver mantle had lost most traces of brown. He had only two adult females. The younger of the two, 'Ridge', had barely reached adulthood and apparently had not yet suckled young. The elder, 'Plush',

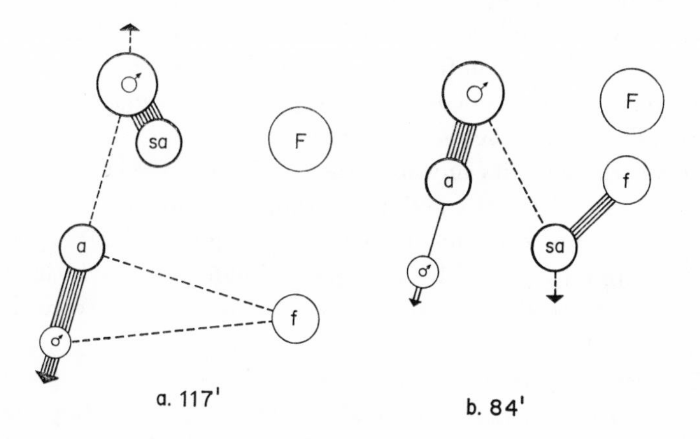

Fig. 37. Sociogram of Rosso's unit, (a) at the sleeping rock and (b) in the waiting area. F = adult follower, f = subadult follower. Male-female distances are 73 and 2 cm in (a), 19 and 78 cm in (b).

had a male infant about 9 months old at the beginning of the 98-day observation period, and apparently became pregnant again near the end of the period. Rosso's unit was accompanied by two followers: A young adult male and a still mantleless subadult. The subadult was occasionally groomed by Plush at the sleeping rock and would play with Plush's infant. The adult follower never came into contact with the unit at the sleeping rock, but would sleep close to it and on route, would follow it constantly. The relationships of the two followers to the unit are described in more detail on pages 54 and 147.

One feature of Rosso's unit is of general significance. The unit's pattern of interactions underwent basic changes each time the unit entered or left the area of the sleeping rock. In or close to the rock, it was always the younger of the two females, Ridge, who groomed Rosso. But when the unit paused at a more distant waiting area, Plush alone did all the grooming, leaving Ridge to turn her attention to the subadult follower (Fig. 37 a and b). The distances between the females and their leader clearly reflect these changes. There was nothing to suggest that Rosso in any way caused them. Plush, the older, was the more dominant of the two females. She often initiated unit movements by leading the way, while Ridge, as a rule, brought up the rear. Plush would respond to Rosso's gaze by calmly moving closer to him, whereas Ridge's response was to run to Rosso while giving staccato-coughs and then present to him. Presumably, the changes in the behavior of the females occurred because Plush, the more dominant female, took her place by Rosso's side in less familiar territory, but relinquished it to Ridge at the sleeping rock in favor of a more comfortable ledge. Regardless of how this is interpreted, the fact remains that social relations in the unit may, superficially at least, be reversed by location.

From early maturity onwards, the behavior of the leader develops along two gradients: His activity within the unit declines steadily, while, at the same time, his influence on troop movements increases. The differing roles played by younger and older leaders during troop movement, especially Rosso's great influence in this regard, will become clear later.

The decreasing activity of aging males as unit leaders is paralleled by an increasing tolerance towards their females. The distance during the march between the old leader and his females increases, and the number of females in such a unit is reduced to one or two. Although Rosso kept his females together, he did so solely by glancing at them.

He never resorted to brow-lifting, attacking or neckbiting, as is frequent among younger leaders. His most severe threat was slapping the ground with his hand when his younger follower tried to copulate with one of his females directly in front of him, which represents a clear overstepping of the average leader's tolerance. Whereas all the interactions of the young leader, Guy, with outsiders consisted of threats, seven out of Rosso's eight contacts outside his unit, all with subadult and adult males, consisted of notifying behavior.

'Pater', a leader even older than Rosso, was associated in a temporary two-male troop of twelve with Circum's unit. He was observed during the course of two evenings and a day. Pater was thin and light haired, with a red face showing wrinkles and bags under his eyes, and his movements were slow. He led only one adult female who rarely groomed him, though he often allowed himself to be groomed by the two-year-old male of his unit. Contacts between juvenile and adult males are rare, which points to the low tendency to aggression in the old male. Whereas Circum's four females were never more than seven meters from their leader when the two units were on route together, Pater's female was allowed to eat at a distance of 30 meters from him with impunity. Nevertheless, Pater determined the direction the two units would take.

'Silver', another aged unit leader, also had only one female. She too was allowed to go a considerable distance from her leader. The contrast between his leadership and that of Smoke, a younger male, is seen in the sequence on page 135.

The decrease in the number of females belonging to aging males is reflected in the following figures. A total of five aged leaders were observed, each in a two-male association with a leader in his early prime. The five young leaders had a total of 18 females; the five old males only had 6. Nevertheless, in each case it was the old male who determined the direction the two units took. From other observations (p. 102) it might be concluded that older males lose their females mainly in fights. Their unrivaled influence in directing troop movement, however, demonstrates a superiority of rank which makes this conclusion improbable. A more likely possibility, suggested by the tolerance of the older males toward straying females, is that the older males gradually and without fighting release the females from their control.

d. The Unit of Mothers

In the types of units thus far described we never encountered more than five females. The frequency diagram on page 31 shows, however, that there are units having 7 to 10 females, numbers well outside the main frequency curve. These units are in many respects unusual and cannot be included in the genesis of the average unit. There was only one such unit in the White Rock troop. It was studied for a period of 55 days.

Sweep (Fig. 38) was a young adult with a pure brown mantle. He was followed by nine females among which were two juveniles, 2 and 3 years of age. Not less than six of his seven subadult and adult females had small black infants, of which half were born during the observation period. The unit's loose cohesiveness was in no way strengthened by its leader. Even on the sleeping rock itself the females remained as much as 15 meters behind their male and frequently allowed visual contact with him to be broken. Only when a strange male sat down too close to them or when Sweep moved on, would they move closer. Social interactions among the members of the unit were extremely infrequent. Now and then a female would present to the leader and groom him briefly, or two females would groom one another for a few minutes. Low frequency of social interactions, however, is typical of most mothers with small infants.

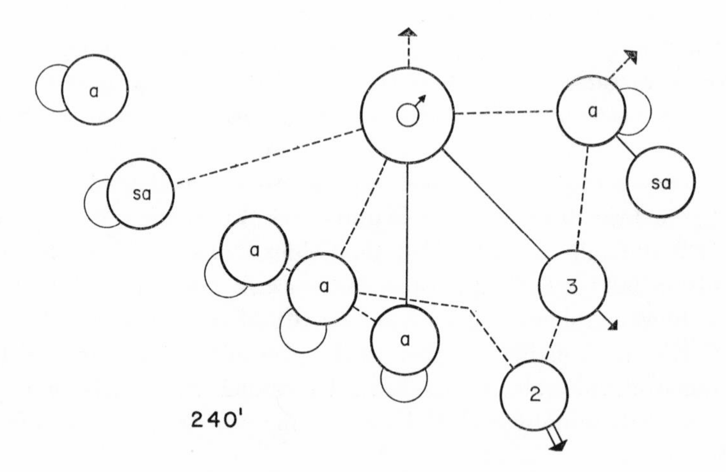

Fig. 38. Sociogram of the mothers' unit (Sweep). Male-female distances are 105, 85, 54, 52, 64, 100, 67, 79 and 47 cm.

More puzzling, therefore, is the leader's passivity in not initiating a single interaction with his females. Had Sweep possessed only his two juvenile females, the older of which was sexually mature, then his unit would have been typical of Sweep's age class. But while such a leader would energetically lead his juvenile females as Guy did, Sweep did not even watch his and would allow them to wander about freely and play with other juveniles. Only the elder occasionally groomed him.

Sweep's neighbors showed a great deal of interest in the unit's black infants. Neighboring females would come and sniff at them. 'The nursery' was often encircled by several young adult males who had no units of their own; some of them even sat amongst the females themselves, which would have induced another leader to violent herding behavior. The leaders of neighboring units often spent time looking at the infants. On two occasions one of the younger male onlookers tried to kidnap an infant. One of them was pursued by Sweep until he released the infant.

Membership in Sweep's unit was not as constant as in other one-male units. On the second day of observation one of the females was missing. A month later she was again observed among the unit together with a new-born infant. On the morning of the 27th observation day another female was missing with her child. On the morning of the same day, still before the break of camp, a fight broke out between Sweep and one of his neighbors.

> After the fight, as the two males sit opposite one another threatening, Sweep looks steadily at one of the other male's females with a black infant who now presses herself to the ground next to the other male. The other male immediately places his hand on her back (behavior when the possession of a female is threatened).

Because of the tall grass I was prevented from identifying the female, but very possibly Sweep's fight was an unsuccessful attempt on his part to recapture his missing female.

The social dynamics that produce a unit of mothers are still not understood. Since mothers with black infants rarely interact with their leaders and are allowed to distance themselves farther than other females without being punished, it is possible that mothers are the first females to leave aging leaders. And since they are drawn to one another by virtue of their infants, they may preferentially join the same unit of a tolerant leader until a more aggressive young male draws them out of this pool of females and incorporates them into his own unit.

Sweep's unit shows that even the least social female class, i.e., mothers immediately before and after they have given birth, is subject to the loose leadership and protection of a particular male and that this relation can exist without any sexual behavior and with almost no direct physical interactions.

4. SOCIOGRAPHIC AVERAGES

The sociograms of the six typical, mature one-male units (with the exclusion of the initial unit and the unit of mothers) are summarized in Table VIII. In spite of the wide range of variation among these units, some common tendencies are evident:

1. The adult members devote only a small portion of their social interactions to animals outside their unit. (If relationships to followers are disregarded, the figure for females is also reduced to 3%).

2. The social activity within a one-male unit moves in two main streams; namely, between the male and his females and between these and their young. This star-shaped pattern, the center of which is the male, keeps the unit together. In most of the units, other

Table VIII. Mean percents of observation minutes spent in social interactions by the members of 6 typical one-male units. The mean duration of observation is 238 min. per individual, distributed over an average of 2 months.

	Interactions of the		
	leader	female	infant or juvenile
a) *within the unit:*			
with leader	—	19	0.9
with females	52	7	mother: 28 others: 0.3
with infants and juveniles	0.2	own: 18 others: 0.2	10
b) *with animals outside the unit,* including followers	2.2	5.4	?
Total	54.4	49.6	?

possible interactions, as those between the leader and juveniles, or between females and females, are not more frequent than contacts with strangers.

5. SUMMARY I: THE ONE-MALE UNIT

In a survey following one fourth of the west-eastern diameter of the species' range, the hamadryas population was found to be fragmented into units of one male and several females. These units associate with each other and with single males in larger bodies, or troops. In a sample studied more closely, no females were found to live outside such one-male units. In contrast, about 20% of the adult males had no unit to lead.

The intensive study on two troops has shown that hamadryas males, unlike the males of other baboon species, do not share their females. Instead of fleetingly consorting with any obtainable female in oestrus, the hamadryas male leads and defends a stable set of the troop's females. Whereas savanna baboons compete for a female only at times when she is sexually receptive, the hamadryas male appropriates and defends females regardless of their momentary sexual condition. When he fights over a female, he fights over her permanent possession.

Apparently, the hamadryas version of the one-male unit is imposed on the females by the males. Whereas females do not reject a chance copulation with an immature male outside their unit, no adult male was seen to seek sexual interaction with females other than his own. Also, the female is more likely to stray away from her male than vice versa. By their behavior, the males constantly segregate the units in the crowd of the troop. Paradoxically, they do so by threatening and attacking their females. By timing the attacks to the female's being too far away, the unit leader reverses the usual effect of aggression into an escape *toward* the aggressor, i.e., toward himself. Generally, such 'reflected escape' is one of the means by which a social species may prevent aggression from disintegrating the group (CHANCE, 1966). In the hamadryas one-male unit, however, aggression and reflected escape even serve to *strengthen* the cohesion.

There are three situations which may provoke the male's attack against one of his females: She may be metrically too far away from him; she may allow an outsider to settle between her and the leader;

Table IX. Reconstructed life cycle of the one-male unit in terms of the leader's behavior, including the pre-unit forms of kidnapping and adoption. The frequency and intensity of behavior is represented by the number of dots.

Stage	'Maternal'	Restrictive	Sexual
Kidnapping	• •	•	
Adoption	• •	•	
Initial Unit	• •	• •	
Transitional stage	•	• • •	• •
Early prime		• •	• •
Late prime		•	•

or she may engage in interactions with an outsider. The leader's intolerance of these situations socially isolates the female from baboons outside her unit.

In the various types of one-male units, one recognizes different stages of the average unit's life cycle (Table IX). The tendency of the male to herd his private set of weaker animals develops even before he actually founds his unit: The subadult male tends to kidnap and mother black infants, and the young adult male permanently adopts and mothers juveniles. Eventually, he adopts a juvenile female and, by threatening her whenever she moves away from him, gradually conditions her to follow him. Thus he founds his own one-male unit in its initial form. The male and the juvenile female in the initial unit still simulate the behavior of mother and child. We therefore hypothesized that the male-female bond in the hamadryas baboon may originate from a transferred mother-child relationship (for details, see KUMMER, 1967). As the unit matures, the mother-child behavior gradually disappears. Also, the mother-like manual retrieving of the straying female, typical of the initial unit, is replaced by the ritualized bite on the female's nape or back.

Sexual behavior appears only long after the unit is established and periodically disappears again when the females are in anoestrus or pregnant. However, the lack of sexual behavior does not noticeably affect the unit's stability.

The peak of the unit's size is reached when the leader is in his early prime. In late prime and old age, the male's restrictive behavior becomes less intensive. He now tolerates his females to move farther away from him, and his unit decreases in size. The leader's activity gradually shifts from leading the unit to directing the travel of the troop, as will be described in the second part of the study. The status of a hamadryas male, therefore, cannot be described in terms of a dominance hierarchy to which all male functions are related. Instead, there are two overlapping roles, related to age: That of the young unit leader and breeder, and that of the old troop leader.

In the section on the male followers of the unit, we have found the first developmental stages of another relationship, that of two males of different age. In the first stages, juvenile and subadult males who associate with a particular unit only engage in furtive sexual and grooming interactions with the unit's females. Later, however, the young adult followers replace these relations by interactions with the leader during the march. The later stages of such associations will be found in the two-male teams, in which two leaders of different age interact in directing their route of travel.

After this manuscript was completed, a review by Bowden (1966) made available some of the unpublished observations made by Tikh in 1950 on the captive hamadryas at the Russian Medico-Biological Station of Sukhumi. The Sukhumi baboons, which in part have lived at the Station for several generations, apparently still organize themselves in one-male units: 'In a compound containing more than one group, (i.e. one-male unit, author's comment), the alpha-males are observed to prevent members of their own groups from mingling with those of other groups in a manner not dissimilar to a mother's preventing her young infant from mingling with other animals within the group.' As to the origin of the one-male unit, Tikh and I independently arrived at the same hypothesis. 'Tikh suggests that the protective function of the older males... may have a common origin with the protective components of mothering behavior' (Bowden, 1966). 'The genesis of the one-male units suggests that a transferred mother-infant relation was an important root of their evolution' (Kummer, 1967). This hypothesis must of course be limited to the hamadryas baboon. The structurally similar one-male units of the geladas and the patas monkeys may be based on very different motivations.

V. AFFINITIES BETWEEN THE SEX-AGE CLASSES

In the previous chapters we relied heavily upon our close study on the White Rock troop. We now return to the level of the broad sample with three aims in mind:

(1) To check whether some of the findings on the identified one-male units can be confirmed quantitatively in a broader section of the population,

(2) To investigate the grouping tendencies of the males outside the one-male units,

(3) To develop some quantitative methods for measuring inter-class affinities, that can be used not only on hamadryas, but also (by independent workers interested in comparative data) on different species.

Clearly, there is no single method that would accurately describe 'social affinity' between classes. We therefore used three measures: Class association in detached parties; spatial arrangement of classes within a party; and the frequency of social interactions between classes. In order to make the three approaches as independent as possible, the data for each approach were collected separately. Furthermore, all data for the first and second approach were collected by myself, whereas Kurt collected the data for the third method. The procedures and results are given separately for each method; the interpretation will then summarize and compare the results.

1. METHODS

In the first method we assessed inter-class affinities by checking the composition of detached parties within or near the troop. The criterion for deciding what was a 'detached party' was the same as that used in the study of one-male unit composition (p. 30). Data were collected in the first 3 months of the field study, mainly during the daily travel.

Whenever we saw a number of baboons that had separated themselves from the others according to the criterion, we recorded their number and the sex and age of each individual. Detached parties containing more than two adult males were excluded from the data. The other parties were classified according to their combinations of the three main classes of animals, i.e. males (adult and subadult), females (adult and subadult), and young (less than 3.5 years old). Table X lists the various types of parties that occurred, and their mean composition. For example, parties consisting of males only (type G) contained an average of 1.65 adult males and 0.35 subadult males. The frequencies (n) of the various types of parties are no exact measure of their actual frequencies, since sampling errors may have occurred.

In the second method, social affinities were examined by measuring the spatial arrangement of sex-age classes within the resting troop. I selected a 'subject individual' at random, and determined who were his three nearest 'neighbors' (Table XI). For each class of animals, 20 to 50 subject individuals from 5 different troops were thus studied. For each subject individual I recorded the sexes and ages of the three baboons nearest to him 10 seconds after the choice. The chance that a subject animal was chosen twice is negligible.

If baboons would mix at random in the resting troop, i.e. ignoring class distinctions, the proportion of times each class appears as a neighbor would be based simply on the proportion of such animals within the population (Table VII, p. 27). Differences between chance expected and actual occurrence for a given pair of classes can be assumed to indicate positive or negative affinity between these classes. One must only keep in mind that the frequencies in Table XI are not completely independent of each other. However, the following steps were taken to increase independence: The data were collected on several troops; the recording was spread out over a period of several months; and on a given day, I observed only one subject class in a given troop, taking care to select the individuals from different parts of the troop.

The results were analyzed in three ways. First, the chi-square values in the last row of Table XI provide a measure of the degree to which the three nearest neighbors of each subject class are selected on a non-random basis.

To determine the expected frequencies, the total number of neighbors recorded in each row was split up according to the percentages in Table VII. Although these figures (Table XI, in parentheses) are merely an estimate of the true expectation, it is reasonable to use them as expected frequencies in the chi-square test since they are based on a much larger sample than the observed frequencies. With 10 degrees of freedom, chi-square values above 19 indicate that the observed composition was non-random at the 0.05 level of significance; a chi-square of 30 corresponds to a P of 0.001.

Secondly, we would like to know which specific classes appear significantly more or less often than expected as neighbors of a subject class.

This was determined by a chi-square test using as observed frequencies a) the occurrence of one class, and b) the total occurrence of all other classes among the

Table X. Mean age-sex composition of different types of detached parties. Figures give numbers of individuals, n is the number of observed parties.

Type A: Male-female-young parties, n = 44

Males		Females	Young				Total
adult	subad.		3 yr.	2 yr.	1 yr.	black	
1.3	0.3	2.3	0.4	0.4	0.8	0.1	5.6

Type B: Male-female parties, n = 19

Males		Females	Total
adult	subad.		
1.0	0.5	2.4	3.9

Type C: Male-young parties, all young are females, n = 13

Males		Female young				Total
adult	subad.	3 yr.	2 yr.	1 yr.	black	
1.0	0	0.5	0.4	0.4	0	2.3

Type D: Male-young parties, all young are male, n = 8

Males		Male young				Total
adult	subad.	3 yr.	2 yr.	1 yr.	black	
0.4	1.0	0.6	0.7	0.9	0	3.6

Type E: Male-young parties, young of both sexes, n = 4

Males		Male young				Female young				Total
adult	subad.	3 yr.	2 yr.	1 yr.	black	3 yr.	2 yr.	1 yr.	black	
0.7	1.5	1.2	1.0	1.5	0.5	0.25	0.75	1.0	0.25	8.65

Table X. continued

Type F: Female-young parties, n = 4

Females	Young				Total
	3 yr.	2 yr.	1 yr.	black	
2.0	0.25	0	1.0	0.25	3.5

Type G: Male parties, n = 17

Males		Total
adult	subad.	
1.65	0.35	2.0

Type H: Female parties: none observed

Type I: Young parties, n = 9

Male young				Female young				Total
3 yr.	2 yr.	1 yr.	black	3 yr.	2 yr.	1 yr.	black	
0.9	1.8	2.3	0.4	0	0	0.6	0.1	6.1

neighbors of the subject class. The expected frequencies were again calculated from Table VII. The resulting statements of significance are indicated by asterisks in Table XI.

We can further ask whether a class appears equally often as closest, second and third neighbors of the subject class, and thus investigate an affinity's spatial gradient. For example, adult males are found 13 times as the nearest neighbors of adult oestrous females but only 6 and 5 times as second and third neighbors, respectively. Chance occurrence would be 5 for each of the three positions. This significant gradient disappears when the females go into anoestrus.

To test the significance of such gradients, I compared the observed frequencies of the neighbor class in the three positions with the frequencies of the same number of neighbors if they had been evenly distributed over the three positions, using the

Table XI. Frequencies with which sex-age classes appeared among the three nearest neighbors of males and females of various ages. Each cell shows the observed frequency and, in parentheses, the frequency that would be expected if the classes were mixed at random. Single and double asterisks indicate that observed and expected frequencies differ significantly at the 0.05 or at the 0.01 level of significance, respectively. > indicates a significant (0.05) decrease of frequency from the nearest to the farthest of the three positions.

Subject individuals by class

Class of neighbors	*Males* adult	subadult	3 years	1 year	*Females* adult anoestrus	adult oestrus	3 years anoestrus	3 years oestrus	1 year
Males									
Adult	19** (32)	40 (30)	17 (22)	18 (24)	31 (31)	24**> (15)	25> (19)	23 (20)	30<> (31)
Subadult	12 (8)	13 (8)	7 (6)	6 (6)	5 (8)	5 (4)	4 (5)	3 (5)	8 (8)
3 years	3 (6)	8 (6)	10** (4)	9** (5)	3 (6)	1 (3)	5 (4)	5 (4)	5 (6)
2 years	1* (7)	12* (6)	16** (5)	10* (5)	7 (7)	1 (3)	3 (4)	9*> (4)	7 (7)
1 year	4 (9)	6 (9)	11* (6)	10 (7)	8 (9)	1 (4)	3 (5)	5 (6)	11 (9)
Black	1 (3)	0 (3)	1 (2)	2 (2)	3 (3)	2 (1)	2 (2)	2 (2)	3 (3)
Females									
Adult and subadult	54 (45)	42 (43)	22 (30)	28 (34)	36 (44)	21 (21)	23 (27)	27 (28)	49 (44)
3 years	14* (8)	1* (7)	2 (5)	3 (6)	8 (8)	4 (4)	6 (5)	4 (5)	5 (8)
2 years	16** (8)	8 (8)	4 (6)	5 (6)	8 (8)	4 (4)	4 (5)	4 (5)	5 (8)
1 year	11 (10)	2** (10)	3 (7)	12 (8)	16 (10)	2 (5)	9 (6)	4 (7)	9 (10)
Black	3 (3)	0 (2)	0 (2)	2 (2)	10**> (2)	1 (1)	0 (2)	1 (2)	3 (2)
Number of subjects	46	44	31	35	45	22	28	29	45
χ^2	33.2	30.5	48.0	20.1	39.7	13.1	7.9	10.0	4.0

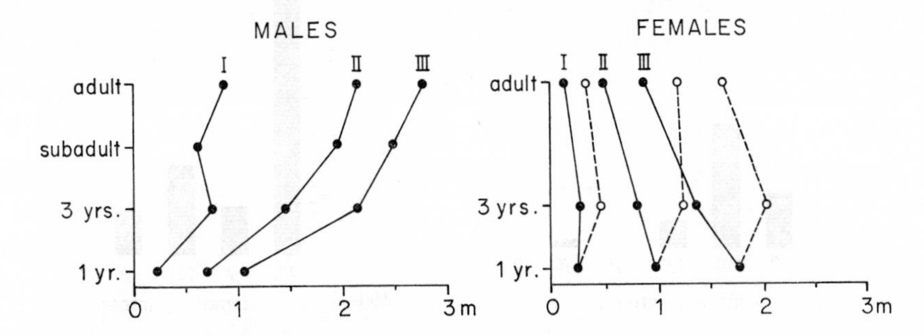

Fig. 39. Age changes of the mean metric distances of males and females from their nearest (I), second (II) and third neighbor (III), regardless of the neighbors' age and sex. Calculated from the sample of Table XI. Empty circles show the increased distances of females in oestrus.

test given by VAN DER WAERDEN[1] (1957, p. 227) for the comparison of probabilities. Significant gradients are marked in Table XI.

The metric distances between the subject animal and his three neighbors regardless of their class are represented in Figure 39.

The third method used the frequency of social interactions between the classes as a measure of affinity.

Kurt chose subject individuals of three classes (94 adult males, 51 adult females in oestrus, and 33 adult females in anoestrus) from 6 resting troops and recorded their social interactions for an average period of 21 minutes. The list on p. 180 shows the behavior patterns that were recorded as 'social interactions'. For each of the three subject classes we determined the number of observation minutes in which the subject individual interacted with each partner class. These figures were then divided by the frequencies of the partner classes in the population. The resulting relative frequencies represent measures of class preference. They are shown in Figure 40, together with the relative frequencies of neighborhood derived in the same way from Table XI. A score of 100 indicates that the partner class was selected for interaction or neighborhood exactly as often as would be expected if the partner's sex and age had no influence on the choice.

2. ONTOGENY OF GROUPING TENDENCIES IN MALES

The results of the three methods will now be compared with those of the close study, tracing the class affinities of each sex from infancy to adulthood.

[1] I am indebted to Dr. VAN DER WAERDEN for his advice on the statistics used in this section.

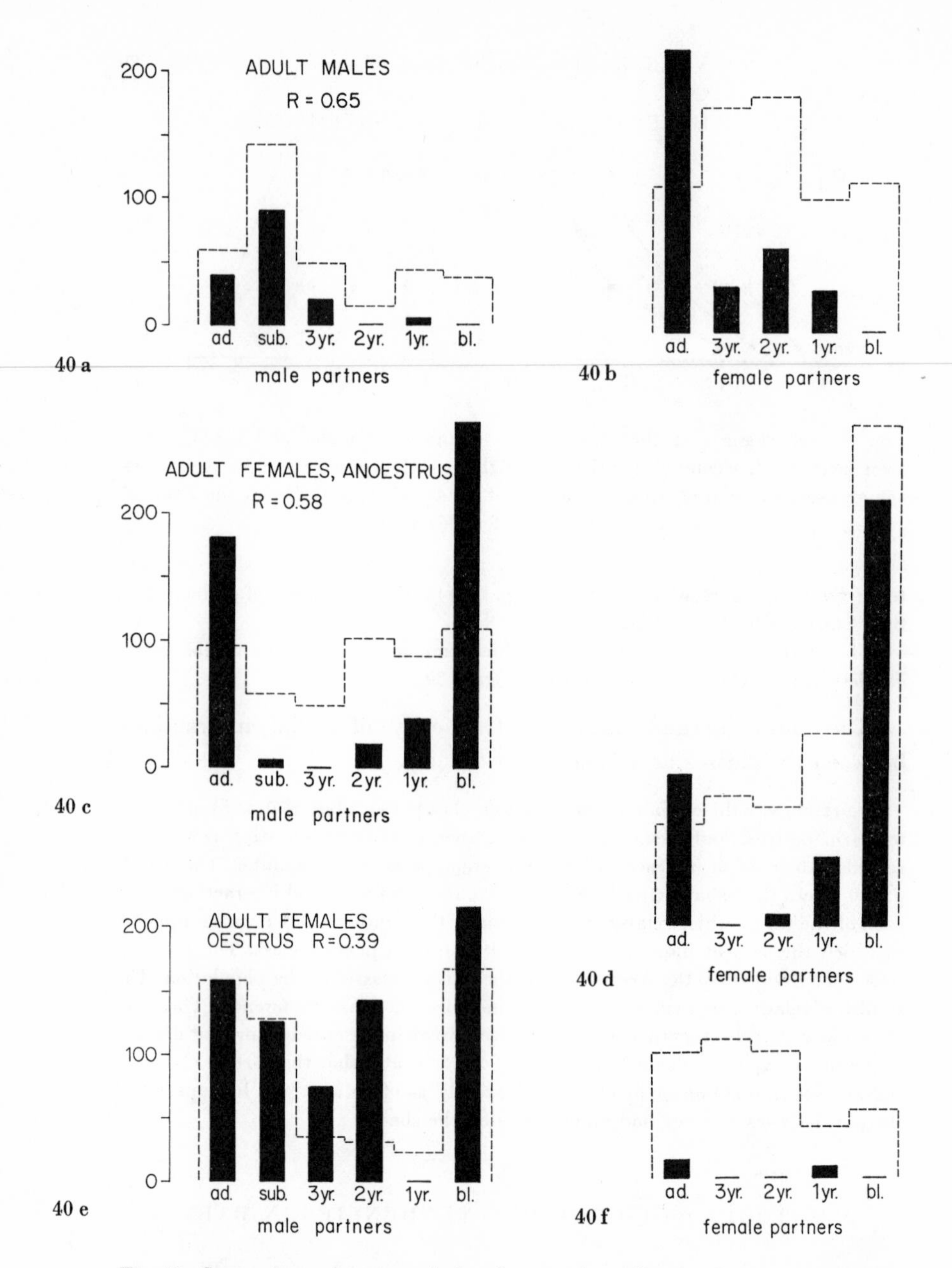

Fig. 40. **Comparison of two methods of measuring affinity: the nearest-neighbor method and the social interaction method. The broken bars indicate the relative frequency with which three subject classes (adult males, adult anoestrous females, and adult oestrous females) were in proximity to various classes of neighbors ('partners'). The black bars indicate the relative frequency with which subject classes engaged in direct social interactions with the partner classes. Scores are percents of chance expectation.**

For *black infants*, the neighbor-class method reveals a striking sex difference: Female infants are the neighbors of adult anoestrous females three times as often as males of their age (Table XI); their social interactions with the adult anoestrous females, however, are hardly more frequent than those of the male infants (Fig. 40). Obviously, the male infants are nursed and groomed by their mothers nearly as often as female infants, but between these contacts, the black female will remain near the mother, while the black male tends to leave her. The analysis of detached parties (Table X) shows where the black males are heading: The 'play groups' of types E and I include more black males than females. The close study did not reveal differences in maternal behavior towards male and female infants. In female rhesus monkeys of the same age, YOUNG and coworkers (1964) have demonstrated an increase in 'rough-and-tumble' play following injections of testosterone propionate into the mother during pregnancy, and HARLOW (1962) found, that at the age of two months, laboratory rhesus males engage in more rough-and-tumble play than females of the same age.

At *one year* of age, the infant male's tendency to join predominantly male play groups is still more prominent (Table XI and detached parties of types E and I). In the second year of life, some of them do not even spend the nights near their mother, but sleep with a few other juvenile and subadult males on a separate ledge. At one rock, such a group had a constant sleeping place, but membership varied slightly from night to night.

The strong affinity among the juvenile males is clearly shown by the neighbor-class method. Significant preferences extending to the third neighbor occur in no less than 5 of the 6 relevant cells in Table XI. The peak of this attraction is reached at two years. At this age, the juvenile males still join the maternal units during travel, but when the troop settles to rest, they assemble near the troop's periphery and start to wrestle and to chase each other. Juvenile females of two years and older tend to join only if the play group builds up around an adult or subadult male (detached parties of types C and E). A single subadult or adult male is often seen forming the center of a play group and watching the wrestling and chasing; sometimes he threatens one of the players or another one screams and flees to him. If such a potential protector is missing, the males greatly outnumber the females: The parties of Type I contained a total of 49 males as against 6 females, all of whom were infants.

Three-year-old males still join the group of juvenile males but play-fighting is less frequent and more inhibited. Instead, three-year-olds often groom each other. Their neighbors are still predominantly males. Their increasing distance from the neighbors (Fig. 39) can be explained by their tendency to sit alone near the troop's periphery. This tendency to separate reaches the peak at *subadulthood.* Older subadult males spend only 18% of their resting minutes in social interactions, which is less than half the 'social time' of any other class (KUMMER and KURT, 1965). Social interactions at the subadult age consist mainly of mutual grooming with a peer, a young adult male, or a very old adult male. Such detached parties (type G) most often include two males who show a marked difference in age. The subadult males' tendency to associate as a follower with a one-male unit is documented by the detached parties of type B.

Before and after the unit-leading age, the *adult males*' grouping tendencies are similar to those of subadult males. 80% of the adult males, however, are unit leaders and, in that case, they associate more closely with females. The total social interactions increase from 18% in the subadult males to 43% in the adult males. Now, for the first time since the male's infancy, female classes appear with more than chance frequency among his neighbors, but also, for the first time, the male becomes separated from members of his own class and from males in general. The long growing period in a predominantly male company has come to an end.

Two characteristics of the initial unit are obvious: The detached parties of type C show again that initial units do not have followers, and the neighbor-class method demonstrates the exaggerated spatial coherence between juvenile females and adult males. The composition of the one-male units in general has already been described by the analysis of detached parties on page 30.

3. ONTOGENY OF GROUPING TENDENCIES IN FEMALES

In contrast to the males, females are not attracted to their own sex at any stage of their ontogeny. Their main affiliations are with adult males. As *black infants* and one-year-olds, they sometimes are found in the males' play groups (detached parties of type I), but beyond this age no female is found in a detached party which does not include a subadult or adult male. A purely female play group was only observed once.

The neighbors of *one-year old* females are essentially a random sample of the population. Their only significant preference is for remaining close to adult males. And even here adult males are selected as second neighbors, rarely as first (or third) neighbors. The same gradient of adult male neighborhood can be seen in one-year-old males, but there it does not reach the 0.05 level of significance. This pattern went unnoticed in the close study. We can conclude from it that the one-year-olds avoid the leader's immediate neighborhood.

At *two and three years* of age, most females are already consorts of an adult male. Figure 40 shows that they stay more immediately close to the unit leader, but spend less time in interactions with him than do adult females. This confirms the findings of the close study in which we saw that juvenile females are easily displaced by their adult competitors when grooming the leader.

Interesting changes occur when the three-year-old females enter into *oestrus*. The predominance of the adult male as first neighbor, so pronounced during anoestrus, disappears (Table XI), the average distance from all three neighbors is increased (Fig. 39), and two-year-old males emerge significantly as first neighbors. This confirms the observations of the close study where juvenile oestrous females were seen leaving the unit and entering sexual relations with young juvenile males somewhere out of the leader's sight.

Adult females live in a neighborhood of little profile. Since most of them share their leader with other females and since they no longer try to stay very close to him at all times, the neighbor-class method gives no indication of the one-male unit phenomenon. The interaction method (Fig. 40), however, shows the predominance of male-female contacts in adult life. Only between anoestrous females and black infants do we find more frequent interactions. Like juvenile females, adult females are seldom found in a party where there is no adult male (types F and H).

Oestrus again has marked effects on grouping tendencies. Like the juvenile, the adult female now keeps significantly farther away from her neighbors (Fig. 39). She even gains more spatial independence from her leader (p. 43). Nevertheless, the adult male is now her most frequent first neighbor since the competition with other females is reduced. The total frequency of social interactions is reduced from 49% to 39% ($p < 0.05$) because interactions with female infants are almost entirely lacking (Fig. 40) and because the sexual contacts with the leader are correlated with a reduction in grooming. During oestrus, the adult female sharply reduces her interactions with all females and increases her interactions with nearly all nonadult classes of males. According to the close study, the female in oestrus is groomed by subadult and even by infantile males and attempts to copulate with them behind the leader's back. In contrast to the leaders who do not approach females in oestrus other than their own, the subadult and infantile males are mobile enough to reach a high relative frequency of interactions. In absolute figures, adult males have only 3 times as many interactions as subadult males and eight times as many as male infants. For unknown reasons, juvenile males are not found near the adult females in oestrus. At this age, they consort with females of their own age.

4. SEX DIFFERENCES THAT ARE INDEPENDENT OF AGE

The analysis of affinities in the broad sample has confirmed many of the observations made during the close study. In addition, it shows some general differences between males and females in grouping tendencies that went unnoticed in the close study. First, let us consider the overall tendency toward a non-random selection of neighbors. As the chi-square row of Table XI indicates, all ages of males show significant (19 or higher) values here; whereas only one class of females does: Adults in anoestrus. This class, which includes nursing mothers, derives its extremely high chi-square from the close association of mothers with their female infants. If we do not score infants as neighbors to their mothers, the overall chi-square for this class is an insignificant 9.6. Males of all ages, then, are found in surroundings of more specific composition than females.

Secondly, a significant decrease of a class from the nearest to the third position of neighborhood is typical of females. Not one such gradient is found among the neighbors of males. This means that females of all ages tend to associate with one and only one partner at a time. Beyond this closest animal, the sex and age of their neighbors is more accidental. Males, in contrast, tend to live in a wider field of selected neighbors, extending at least to the three animals next to them. The adult female's prevailing orientation to one male or one infant within the one-male unit, and the adult male's place in the network of the multi-male troop, thus are apparent as general grouping tendencies already in the juveniles.

Thirdly, Figure 39 shows that the distances kept from the neighbors decrease with age in the female, while they increase in the male. Both changes are significant at the 0.01 level. As one-year-olds, the females avoid the dense play groups of the males. Instead, they explore the ground and play more often in groups of two. Later on, they are crowded together by the herding behavior of their leaders, whereas the males prefer the scattered peripheral sections of the troop. Only as subadults and adults, more and more males reenter the crowded center as followers and leaders, while the remaining ones become even more peripheral.

5. EVALUATION OF METHODS

Social relations are usually described by the frequency and sequence of different categories of behavior patterns. For close studies this is the indicated method, but for the study of broad samples it is impracticable, since interactions are infrequent and waiting for them to occur takes too much time.

Now, CARPENTER (1964) has suggested that interindividual distances be used as a measure of social affinities. More specifically he assumed that proximity is an indication of 'positive interactional motivation' between two animals. HEDIGER, in his work on critical distance and flight distance (1934, 1951) has shown that a specific response tends to occur at a specific distance. In using the metric distances between two animals as an indicator of their relationship, we assume that it is in effect a lasting summary of their previous interactions. If this is true, distance patterns are ideal measures for the analysis of large samples. In recent field studies, troop compositions were indeed more often determined by criteria of spatial detachment than by criteria of interaction. Unfortunately, however, we are still far from knowing the exact correlations between distance patterns and the kind and frequency of interactions, and broad samples, therefore, still require observations of actual interactions as controls.

The evaluation of our data allows two comments on these general statements. First, the neighbor-class method yielded valuable results when the *relative* proximity between the sex-age classes was used as a measure. In the same sample the absolute metric distances between two classes were so variable that no significant differences could be found. (The only significant result is that the greatest distance between any two classes is the one kept by adult males among themselves.) Whether two animals are separated by one or two meters of distance seems to matter less than whether they are separated by one or two other animals. The close study has shown this clearly for the one-male unit: A female separated from her male by 5 meters of open space will usually go unpunished, but if outsiders move into the gap, the female will almost certainly draw a neck bite even if much closer to the leader. The importance of relative rather than absolute distance is again apparent among one-year-olds who keep one other animal between an adult male and themselves.

Secondly, the data permit us to compare the frequency of neighborhood (i.e., the occurrence of one class among the three nearest neighbors of another) with the frequency of actual social interactions (Fig. 40). It should be noted that among these social interactions, positive exchanges such as grooming, sexual and play patterns by far outnumber the aggressive and escape patterns (KUMMER and KURT, 1965). As might be expected, actual interactions are less accidental than neighborhood, and therefore, the profile of their frequency is more pronounced. In order to determine the degree of correlation between the spatial and the interactional measure of social relations, Spearman's rank order correlation (R) has been calculated for the three adult classes (Fig. 40). It is at once apparent that the degree of correlation varies according to the sex-age classes involved. For the adult males' relation with all classes of *males* (excluding the female partners), the correlation is almost perfect ($R = 0.97$), i.e., chances of being close have the same distribution as chances of engaging in interaction. If female partners are included, the correlation drops to an R of 0.65, and the R values of the two female classes are even lower. This seems at first to be due to the fact that the class of second and third neighbors is irrelevant for females. However, if only the closest neighbors are taken into account, the contrast between the three correlation indices in Figure 40 is even greater.

An explanation for such marked differences between frequency of interaction and frequency of neighborhood is found in the close study. Where such a difference occurs, a third class usually affects the relations of the two classes under study: For instance, the relation between adult males and adult females is marked by frequent interactions but infrequent neighborhood. In the one-male unit, the female is frequently separated from the leader by other unit members, but once at his side, she makes use of the chance and grooms him. As the figure shows, juvenile females hardly have a chance for interactions with the leader in this competition, although they stay more often close to him than do the adult females. Tripartite relations also appear in the relations of adult females in oestrus. There, the intolerant leader occupies the space around the female; therefore, the younger males generally keep away, but when they have a chance to furtively approach the female, they will almost certainly interact with her. More thorough investigations on the correlation between the two measures may show that a low correlation index indicates a competition of two classes (or animals) for interaction or closeness with a third one.

VI. ORGANIZATION OF THE TROOP

In the remaining chapters we shall focus our attention on the relationship *between* the one-male units and on their tendency to organize in larger associations, i.e. in two-male teams, bands and troops. We shall first present evidence on the structure of troops and bands. A descriptive and an 'experimental' section will then deal with the problem of tolerance between units in relation to their mutual familiarity. In the later sections, we will describe the coordination of travel among two and more one-male units.

1. INSTABILITY OF THE TROOP

We have been using the expression 'troop' for any gathering of animals spending a given night on the same sleeping rock. The number of animals on a particular rock can fluctuate from night to night. A sleeping rock regularly used by over 100 animals may even remain completely deserted for a night, and be occupied again for weeks on end by troops of varying sizes. In 24 counts, we found an average of 121 animals on each of the four regularly used rocks in the vicinity of Erer-Gota. Between two subsequent counts there was an average difference of 31, i.e., 26%. The corresponding figure for the seasonally used Ravine Rock is 48% (21 counts). On two occasions, a week apart, all animals on 6 neighboring sleeping rocks were counted. The counts were taken in the evening, upon the troops' entrance into the rocks, or the following morning, during their departure. None of the troop sizes found in the first count was found again in the second, with one exception where a number found at one rock during the first count was found on another a week later.

Even when counting error is taken into consideration, the results of the two counts clearly indicate a wide variation of troop size. If troops had a constant number of members, the difference between our two counts could be explained only by an almost total substitution of the area's troops by troops from adjacent areas during the week between the counts. This is highly improbable, since one of the six rocks counted was the White Rock where we found the majority of identified baboons on each of about 150 evenings.

Sleeping societies, therefore, are no stable social units; they must consist of separable components which are usually, but not always, to be found at the same sleeping rock. Direct evidence for this is given us by the following:

A troop of 82 animals having spent the night on Ravine Rock, in the mountains, made their way north to the Rotten Rock the following day. Here they spent the night with 40 animals arriving from another direction.

One late afternoon, about 16.30 h, 67 animals were sitting in their ledges on Rotten Rock. Apart from them, on a hill 200 meters to the south, another 65 animals were waiting. At 16.50 h, the 65 animals began to move southwards and after a forced march over some eleven kilometers arrived at their home rock in the mountains just before dark. The liquid feces along their trail gave evidence of the state of their excitement. Presumably the two troops had arrived at the Rotten Rock independently. No contact between them was observed. Moreover, there was no lack of space on the Rotten Rock since 160 animals had previously been observed there.

Apparently, not all the troops in a particular area have a relationship permitting a joint overnight stay.

The smallest independent entity within the troop may well be the one-male unit. Its independence, nevertheless, appears to be limited. Despite the fact that single one-male units were twice observed alone on route, the smallest troops ever observed consisted of two such units (four observations). The section on troop movement will show that travelling parties of two one-male units are frequently formed and that these will stay close together at least for some days.

2. TRAVELLING AND FIGHTING BANDS

There are, however, several indications that between the two-male units and the troop, other entities of medium size exist. These entities, which we called bands, are not apparent under usual circumstances. At first we became aware that if counts of troops are repeated sufficiently often the same figures may appear again. In 50 counts made on Ravine Rock and the neighboring Rotten Rock,

41a

41b

Fig. 41. Phases of large scale fights. (a) Artificial feeding causes the units to crowd and intermingle. Leaders first respond by biting their females (left) or crouching over them (center). (b) Females start to line up in the 'attack shadow' of their

41c

41d

leaders (right). (c) Star pattern of three opponent units. Mounting 'during tension' (Fig. 5b in KUMMER and KURT, 1963). (d) The male on the right snaps at a female. Her leader (left) responds by threatening the aggressor.

the figures 62 or 63 were found four times, 68 twice and 81 or 82 five times. Furthermore, on several occasions we noticed an old male, Rosso, travelling with a column of some 43 animals apart from the rest of the troop. (Counts on three different days showed 42, 43, and 43 animals.) On route we often encountered large, isolated parties of a troop who would arrive at the sleeping rock independently of each other. Twelve out of the nineteen counts of such bands in the Erer Gota area came to between 30 and 60 individuals. The White Rock troop consisted of two bands, a smaller one, including Rosso, which always slept on the right half of the rock, and a larger one, including all other identified units, occupying the left half. The two bands usually arrived at the rock at different times in the evening.

Other indications are to be found in the general battles within a troop which, of all our observations, were the most impressive. One spontaneous battle of this sort was witnessed, as well as four others which we induced by emptying corn in a pile near the sleeping rock. Two of these induced battles were filmed and analyzed; one at the White Rock, the other at the rock belonging to a troop near Diredawa. All four induced battles occurred in three successive phases. In the first, 10 to 20 animals would gather near the pile of corn and begin to eat. In this way a number of one-male units were forced into so small a space that, in violation of the 'rules', they were compelled to intermingle. At first the leaders responded to this by giving their females bites on the neck or back (Fig. 41 a).

After about 10 minutes the second phase would begin and the first duels between unit leaders start. During pauses in the fight, the leaders would lay themselves bellywards over one of their females (Fig. 41 b), and threaten one another by raising their brows and rhythmically pumping their cheeks. Females would tend to line up in their males' 'attack shadow' (Fig. 41 c). Now and then a densely packed one-male unit would follow its leader away from the place of fighting in a hasty retreat. Evidently, the prolonged intermingling led to an increasing insecurity among the leaders over the possession of their females. Soon the fight began to spread beyond the feeding site to other parts of the troop. Apparently the fight was less about food than about the integrity of the intermingling one-male units. Our films show that individual males persistently attacked particular females of other leaders (Fig. 41 d). Sections of the troop just arriving at the sleeping rock would also engage in the battle

41e

41f

Fig. 41. (e) The band in front is closing ranks against the band on the left and rear. (f) The rear band is withdrawing up the river bank. Fighting is going on (right of center).

even though only a few kernels of corn remained. The ceaseless screeching of the females and the two-syllable bark of the males became quite deafening.

In the third phase, some 30 minutes after the beginning of the battle, the troop would begin to disintegrate. Two or more single bands of 20 to 90 individuals would form; several of them then occupied the neighboring hills and threatened the band still at the feeding site. At intervals, one of these peripheral bands would run against the feeding band and would either drive them away or be driven away by them (Fig. 41 e, f). In typical encounters, the adult males of each band would form the main front; though ahead of them the subadult males would stand and alternately threaten their opponents and look back at their own adult males. The females usually remained behind the front line, though occasionally they too would dash out and come quite close to the opposing males. Infants were carried by the adults. Actual physical contact between opposing bands was rare, since the attack of one band was synchronized with the retreat of the other. This has already been seen to occur in duels between individual males. Despite the enormous expenditure of energy in screaming and chasing about, no injury beyond a few cuts and scratches was ever observed. The attacks and withdrawals continued until onset of darkness when band after band returned to the sleeping rock.

The spontaneous fight occurred one evening at the White Rock when, for unknown reasons, a large section of the neighboring Red Rock troop attempted to spend the night on the White Rock. Essentially, the battle took place in the same three phases showing that the food itself had little to do with the pattern of battles.

The Red Rock troop had split up a few hundred meters in front of its rock on its way home. One section of it moved toward the Red Rock; the other toward the White Rock. At 17.30 h I found the White Rock overcrowded but quiet. At 17.40 h, a clamor began from the right of the rock. At this the animals on the left started to flee from their ledges returning only after some 10 minutes. At 17.57 h a fight involving first two, then three males broke out on the right. Normally a troop as a whole does not respond to such fights. This time, however, the onlookers reacted with screams and barks and finally all the animals left their ledges. During the course of the local fight, one of the fighting males fell from the cliff; five other males immediately sprang down to him and the fight was continued in the river bed. A minute later several males in the river bed were seen covering their females as other males attacked them and reached out at the females. Eight males were now near the battle ground; above them the last of the animals were leaving the rock along the most frequently used route.

At 18.20 h the crowd had fallen apart into several isolated bands, each sitting on one of several hills in a semicircle some 200 meters to the north of the rock. Some of the parties were quite calm, while in others fighting was still going on. Individual one-male units split off from the rest as though in flight; all over liquid feces could be seen.

At 18.45 h twilight began to fall. Closed parties of 33, 28, 30, and 93 animals began gradually to leave the hills and move toward the cliff. The last band was once again attacked by the preceding one and driven back. At this, one of the females from the attacking band fled to the other and was received into the arms of one of its males who was probably her original leader. At 19.05 h darkness had set in and the rock was completely occupied. The noise and barking which had lasted for over an hour finally stopped.

The next morning the screaming began again in the dark. At dawn, about 05.45 h, the White Rock was already empty. To the north, on a mountain some 1½ kilometers away, two parties were sitting at a great distance from one another. On the way to the Red Rock, about 2 kilometers off, I encountered another 120 animals in restless detached parties, at first moving about in different directions, but finally coming together in a southward march. At 07.00 h the Red Rock itself was also unoccupied, but was partially covered by fresh liquid dung. Apparently, a new conflict had broken out before dawn on the White Rock, after which the Red Rock baboons shortly returned to their own rock, while the White Rock troop withdrew to the mountains. That evening, the baboons at the White Rock were asleep more than an hour earlier than usual.

Of interest here is the phenomenon of troop disintegration, which was common to all five battles. Prolonged fighting over the females caused the troops to split into bands which then fought each other. Even in the battle following the Red Rock troop's attempt to settle on the White Rock, it was not the two troops that fought each other, but again the battle was fought among smaller parties. Before and between the attacks, the bands ran away from each other. (The tendency of parties to leave the rock in critical situations sometimes became evident even during duels between individuals.) This raises doubts about the nature of the unstable troop as a social unit. Probably, the troop is merely a loose association of several social units, or bands, which, because of the scarcity of sleeping cliffs, are brought together into sleeping aggregations, or troops. Apparently, however, not any two bands of an area have a relationship that permits them to sleep on the same rock. The prerequisite for sharing a rock was not fulfilled by the two bands that both attempted to spend the night on the Rotten Rock (p. 99); in this case one of the bands had to move to another rock. Neither was it fulfilled by the bands of the Red and White Rock troops so that a spontaneous battle resulted.

It is important to note that a battle gradually develops from skirmishes within the bands into a fight between bands. This suggests that the aggression within the band is gradually redirected against outsiders, i.e., against other bands. As LORENZ (1963) pointed out, such redirection is one of the important mechanisms that control intra-group aggression and strengthen group cohesion. Two fights provoked in 1964, however, never developed to this third stage.

The sizes of the fighting bands of the disintegrated troops corresponded roughly to the sizes of the bands which split off from one another on route. We shall see later, in the chapter on troop movement, that at the departure of a troop from its sleeping rock, parties of similar size determine the direction of their march independently of each other. It seems more probable that the bands appearing on route, during battle, and during the break of camp are identical and remain constant over longer periods of time than that they are formed *ad hoc* by different animals on each occasion. However, we have as yet no evidence for this assumption.

3. NUMBER OF TROOP MEMBERS CONTACTED BY AN INDIVIDUAL

Apparently, the critical question for an understanding of the hamadryas troop is this: Why at all do one-male units associate with each other, instead of keeping apart like the units of patas monkeys? The fights described in the previous section have shown that the units' coexistence is a potential threat to their integrity. Spatial mingling of units results in competition for females, and in such crises the units tend to flee from each other. What are the factors that counteract such avoidance, and that allow at least certain units or bands to remain together? This question is the more difficult to answer since even within the band, unit leaders carefully keep out of each other's reach. They do not exchange any of the behaviors that generally seem to reinforce associations among primates, such as grooming, sexual behavior, or play. Mutual tolerance and attraction among them may merely depend on communicative interactions over a distance, or on a persisting effect of their relationships before the unit leading stage.

If such 'familiarity' is assumed to be critical for the relation of units and bands, how far does it reach beyond the limits of the one-male unit? Within this complex question we tried to answer two specific ones: With how many troop members does a hamadryas baboon observably interact? And what happens if a complete stranger is released in front of a troop?

The close study has shown that leaders and females spend very little time with troop members outside their one-male unit. How many outsiders were addressed by the unit members is not known, since they were not identified. The only available figure is the number of interactions with outsiders. During the observation periods given in Table VIII the leaders had 3.7 ($\pm$ 0.44), and their females 2.2 ($\pm$ 0.83) such interactions. The average number of partners of course could not exceed these figures. During the troop's decamping, when leaders most often interact with each other, the highest number of other adult males contacted by one of 78 males was 5.

These extremely low figures even for the leaders could be explained in at least two ways: Either a unit leader will contact many more troop members if observed long enough; or the figures are at least of the correct order of magnitude, but familiarity is pre-established by far-ranging contacts among the juvenile and subadult males before they become leaders. We tried to test the second hypothesis by observing 5 males before the unit leading stage and identifying all the partners with which they interacted at the rock. For this purpose, Kurt chose a 3-year-old male, three subadult males of different ages and a single young adult male. He observed each of them in 3 to 4 sessions of one hour each, distributed over an average of 25 days.

It was found that the number of troop members contacted ranged from 1 (by the adult male) to 7 (by one of the subadults) with an average of 4.2 ($\pm$ 1.6). More interesting is the fact that each of the 5 males interacted with a stable set of partners throughout the 25-day period. During the first session, the males contacted a total of 17 partners. In the following sessions only 4 new partners were added. 81% of all partners ever contacted appeared in every session, and during the second half of the 25-day period the partners of the first session were not missing more often than in the first half of the period. This shows that even the unattached young males do not make contacts throughout the troop but have a rather limited circle of partners. Kurt also identified the one-male units which slept closest to each of the 5 males. In 20 of the 24 sessions these units were the same for a given male.

The juvenile and subadult males mostly groomed males about their own age, but they also notified the leaders of neighboring one-male units and played with their male infants. 20 of their 21 partners were males of all age classes.

These preliminary results suggest that the males before the unit leading stage interact with a relatively stable set of peers, which, by its size, might at best be the nucleus of a later band, but not of a troop. If the results can be substantiated, they may help to explain what the large-scale fights have shown, namely, that there are within the troop sets of leaders who stay together and fight as a band even when the lower cohesion of the troop as a whole has yielded.

4. EXPERIMENTAL TRANSPLANTATION OF BABOONS INTO FOREIGN TROOPS

The following transplantation experiments were performed in an attempt to form hypotheses about the role of 'familiarity' among troop members by introducing unfamiliar animals into a troop. Two juvenile yellow baboons raised in captivity were the first animals to be released. Their immediate acceptance by the White Rock troop encouraged the transplantations of wild hamadryas. The hamadryas transplants were captured at the sleeping rock near Diredawa and released at the White Rock (the place of the close study), 37 kilometers away from the place of capture. One of the one-male units was transplanted in the opposite direction. The first of these experiments were performed in the last weeks of our field study in 1961, the second series during the making of a film in 1964.

To capture the animals we placed a box with two wire net walls in front of the 'donor' troop (Fig. 42). This trap was some 2 meters in length and had an entrance of 1×1 m on the narrow side. A trap door at the entrance was operated from a distance of 30 meters by means of a rope. Kernels of corn were used for bait. Although the 100 to 700 animals who would watch the capture would occasionally threaten and dash out at us, they never seriously attacked us. Some of the onlookers of the first capture were captured only some minutes later. The only attempt to free the animals from outside came when two females had been captured without their male; the male tore at the wire net until we approached.

As soon as the box was darkened, the captured animals themselves would stop all attempts to escape. By darkening the box, the animals were not only protected from hurting themselves, but it was possible to use lighter boxes to carry the animals in. The animals would then be driven by jeep to the rock at which they were to be released. They were sprayed with a marking dye manufactured and donated by Geigy, Basle. If the dye is applied immediately before the animals are to be released, the color is transferred to those animals with which the marked ones come into contact. Surprisingly enough, neither the marked animals nor their partners paid any attention to the paint as far as we could see. The animals were released 12 to 36 hours after their capture. During this time, at least some of the animals drank and ate. All transplants were released in the evening, after the

Fig. 42. Our trap at the foot of the sleeping rock at Dire Dawa.

receiver troop had settled on the rock. The transport boxes were opened about 10 meters from the foot of the rock, with the opening directed at the rock. Four of the releases were filmed.

a. Transplantation of Single Females

All of the four single hamadryas females which were transplanted were adult or subadult and at the time of the transplant were in anoestrus. They were captured in Diredawa and on the evening following the capture were released at the foot of the White Rock, at a time when the greater part of White Rock troop had already arrived back at the rock and watched the release. Two of these females were released in 1961, on different nights; two in 1964 on the same evening, one after the other. All four releases took a similar course and can be summarized as follows:

As soon as the box was opened the female would run towards the White Rock and its troop. Each time a male from the troop would come running toward her and take hold of her or give her a neck bite and would then lead her back to his other females on the rock, looking around at her repeatedly as in a following procession. During the first hour, the male would often threaten his new female, in between times, mounting her excitedly and grooming her intensively. The females, themselves, would readily follow the onrushing males, already grooming them after the first few minutes. A day later their behavior was hardly distinguishable from the other females in the unit.

When the captured animals were released, most of the members of the troop would move towards the edge of the cliff and watch as the animals ran from the box. Only once did a leader who had commandeered a female get into a brief fight with one of his neighbors. In general, as soon as the female had been taken over, the troop no longer responded to its new member in any particular way. All four females remained in the respective units of their first males, at least as long as we were able to observe them (9 to 13 days).

The first of the females, 'Green', was a subadult. She was taken with a bite on the neck by 'Rob', a young unit leader. Rob had often been displaced by other unit leaders at the sleeping rock and was certainly not a dominant male. When the box was opened, many of the animals rose and ran towards Green. Nonetheless as soon as Rob had bitten Green on the neck, the others withdrew, and as the two animals made their way back in a following procession to Rob's other female, 'Patch', none of the others interfered. After a short time, Patch's infant began to groom the new female. After a lengthy grooming session with Rob, Green preceded the unit down into the sleeping rock and at dusk was seen grooming Rob together with Patch.

The following evening we released the fully adult female 'Red'. She too was fetched by Rob without interference from other males. This time, however, Rob did not bite her on the neck, rather he simply grasped her sides and stared at her face and then led her back to the rock. At first, Red, Green, and Patch made no contact with each other. Then Patch, some minutes after Red's release, began to threaten Red, in between times rummaging in Green's hair with rapid movements.

The unit was studied closely in the next two weeks (Fig. 43). While the original female, Patch, never became aggressive towards the subadult female, Green, she continued to threaten Red for days (Fig. 44), screaming at her, at the same time grooming Green frantically. Red was clearly the most dominant of the three females and would sit closest to the male. As soon as she would leave her place next to the leader, Patch would run to him and groom him intensively. After a few days Patch ceased to threaten Red, staying more to the side, and once grooming a juvenile female which had newly joined the unit.

In general, rapid excited grooming is directed toward the leader by quarreling females. In this case, Patch apparently redirected this grooming towards Green, since the place near the leader had been taken by Red.

Except for the already mentioned fight, the third release followed the pattern of the others.

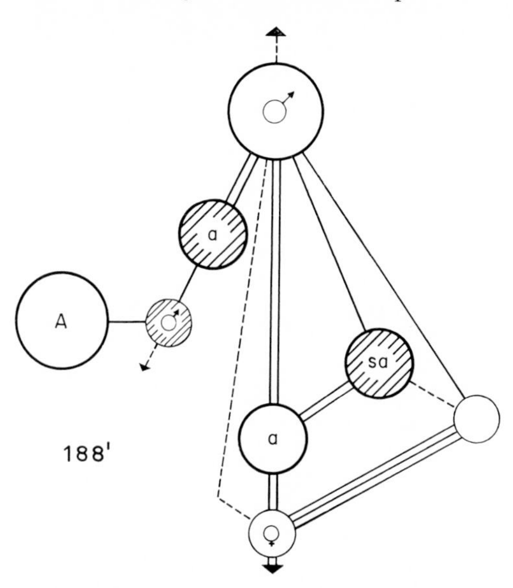

Fig. 43. Sociogram of Rob's unit. Shaded circles represent the implanted females Red (left, with juvenile) and Green. A = Adoptor of Red's juvenile. Empty circle = newly arrived juvenile female. Male-female distances are 21, 103 and 80 cm.

Fig. 44. The transplanted females Green (left, note marking dye) and Red (right) after being accepted in Rob's unit. The original female (center) is threatening Red.

The fourth female also ran to the sleeping rock, at first. However, due to the accidental interference of one of our assistants, she suddenly turned and fled into the savanna. Three or four adult males immediately began to run after her. Dozens of others began to follow them. These, however, soon stopped and sat down in a long line extending from the rock to the pursuers, facing in the direction of the escaping female. Some 15 minutes later I found her about 1 kilometer away. One of the adult males of the White Rock troop was sitting next to her and before dark proceeded to lead her back to the sleeping rock. From then on she became a member of his unit.

b. Transplantation of Entire One-Male Units

Twice, in 1961 and again in 1964, a complete one-male unit was transplanted into a foreign troop. The first time the transplant took place from Diredawa into the White Rock, the second time the other way around. Each unit consisted of an adult male, two adult females and an infant. Each time the spatial relations of the animals as well as their behavior was studied long enough before capturing them until we could say with certainty that a one-male unit was sitting in the trap. Again, all females were in anoestrus at the time.

As soon as the box was opened, the Diredawa unit, led by their male, ran toward the White Rock. But after a few meters the male suddenly stopped, turned about and with his females fled into the savanna. And although the unit was followed by males from the White Rock troop until twilight, it never returned. In order to prevent a similar escape, we did not release the second unit at the edge of the receptor troop, but rather in the middle of the troop. The unit was immediately attacked by the adult males of the troop. In less than an hour the released male lost, in repeated fights, both of his females to native adult males and finally left the troop by himself.

The leader of the first unit appeared to resort to flight at the moment he discovered the foreign troop on the rock. Closely followed by his females, he ran rapidly along the dry riverbed. The one-year-old infant barely managed to hold onto its mother's belly with its hands. Ten meters behind the fleeing unit, several subadult and adult males from the troop were in pursuit. Again, as in the case of the escaping female, the troop had begun to send out a long pseudopod in the direction of the flight. The tip of the pseudopod consisted of males, juvenile to adult, which rapidly disappeared. Later, one-male units also began to wander into the pseudopod, sitting at various distances from the rock. All the animals in the pseudopod were facing the direction of flight, the males calling out 'Bahu' in alarm. Some ten minutes after the release, the young adult male, Mambo, who had earlier adopted a small yellow baboon, turned from the developing pseudopod and came back to the rock. On his back he was carrying the dyed infant of the released unit.

While Kurt remained back at the rock, I followed the pseudopod. The units of the pseudopod were sitting down in increasing distances from one another, on the rises along the path in which the released unit had fled. They were all facing the direction of flight. At about 300 meters away from the rock there were hardly any more females in the pseudopod. At the tip, about 1300 meters away from the rock, and far ahead of the others, an older subadult and two single adult males were sitting on an Acacia. Two hundred meters ahead the foreign unit was just crossing the riverbed which they had been following until now in the unfamiliar terrain. Keeping a good distance, I followed along with them. After another 500 meters they hesitated and retreated 20 meters back toward the rock, and finally disappeared going in their former direction. The three males behind them advanced a bit, but then remained seated, on a tree, continuously giving the 'Bahu' bark (alarm or location of distant baboons). Only at dusk, more than an hour later, did they return to the rock. We never again saw the released unit.

The main part of the pseudopod had already returned to the rock and had released a chorus of contact calls and renewed watching in the flight direction among the remaining animals. On the following morning the troop broke camp earlier than ever before and took, in quick march, the unaccustomed direction in which the released unit had fled. A pseudopod had already formed at daybreak in this direction, which consisted entirely of males.

The second one-male unit was captured at the White Rock and released at the Diredawa Rock. Before we released them we scattered some corn in a radius of 10 meters around the box and in this way we were able to gather a part of the resident troop about the transport box.

As we opened the box from a distance of 40 meters, the unit ran out, led by their leader, and crept into a crevice at the foot of the rock. Several of the troop's adult males followed them, grabbing out at the unit's females, but being held off by the leader's bites. Finally the unit was driven out of the crevice, their leader being involved in heavy battle with resident males, one of which received a lightly bleeding wound near his ear. Whenever he was able to, the foreign male placed himself squarely over his females. But since the mother of the black infant did not stay close to him, it took only a few minutes until she was grabbed by one of the troop's males and driven off. She followed her new leader immediately and he led her together with his other females, a subadult and a three-year-old, to the rock without taking part in the further happenings.

The other female stayed in close body contact with the besieged leader. In the fight that followed with another of the troop's males, he barely ever let go of her and in this way was able to maintain her for more than 10 minutes. After leading her through the troop, he sat down on the plateau above the cliff. As he sat there, continuously holding onto his female, at least 100 of the troop's baboons began to gather about him. While they sat and watched the two, for a few minutes they neither threatened nor attacked them, and limited themselves to watching the pair who sat in the midst of an empty circle about 15 meters in diameter. Later threatening and attacking were again resumed, whereupon the couple withdrew into a field of opuntias.

From the noise, it appeared as though the battle continued there. Ten minutes later the released leader appeared on the edge of the sleeping rock alone. He slowly approached along the edge of the cliff, accompanied, strangely enough, by the entire troop of 350 animals going along with him on all sides. The striking thing about this was that none of the troop threatened him nor in any way made contact with him, but rather would merely follow his gradual shifts very carefully. When he sat, they would also sit and watch him, all the while leaving a very marked area around him open. No other behavior could have so clearly illustrated the foreignness of the stranger.

Two days later we found both of the released females in two different units of the troop. The chests of the leaders of these units were smeared with the red marking dye, indicating that they were the males that had originally taken possession of the females. The black infant seemed to be well. The foreign male, however, despite the striking dye, could not be found and had apparently left the troop.

c. The Fate of the Released Infants

In the experiments described above a total of four brown infants and juveniles had been released. Three of them were separated from their mothers during the course of the experiment and were caught and 'adopted' by three different subadult or young adult males who mostly did not yet lead units of their own; the young animals who would frequently try to escape were each time caught (but never bitten) by these males and were often held tightly in their arms when sitting. On the march, they would be carried on the backs of the males for longer periods than was usual for their age class. All three were rarely seen playing.

The first juvenile, a one-year old male, had been trapped and was released together with the female, Red. The next morning, when the troop broke camp, Red was still carrying it on her back. In the evening, however, it was adopted and carried to the sleeping ledge by a young adult male, Notch, who had arrived a few meters behind Red's unit (Fig. 45). During the rest of the two weeks of our observation period the infant remained with Notch.

The first one-male unit to be released had lost its one-year-old female during the flight. It was brought back onto the sleeping rock by Mambo who carried it on his back (Fig. 46). When it repeatedly tried to flee, Mambo would catch it and bring it back each time without biting it or threatening it. On the day following the release, it was sending out high pitched calls of alarm, as it was held in Notch's arms. A day later it still gave the humming of abandoned infants. Twelve days after the release it was seen playing with the juvenile female in Naso's initial unit.

The third infant, a barely one-year-old female was deliberately set free 15 minutes after its mother (the fourth of the singly transplanted females) had been released. An older subadult male ran towards it, dropped his hindquarters and carried it to the rock. He kept it until darkness fell; but two days later we found it in the one-male unit which had taken up its mother. The subadult male, whose back had been dyed red during the adoption, was sitting next to the unit, looking at the infant. Every attempt by him to approach the infant was thwarted, however, by the threats of the mother's new leader. The subadult male returned the threats and even began a fight, but in the days that followed, was nevertheless unable to win back the infant.

Fig. 45. Notch adopts the juvenile male transplanted with Red. Notch looks at the juvenile male (a), carries it to his ledge (b) and holds it as it tries to climb away (c).

a b

Fig. 46. The female infant lost by the first released unit, and the young adult male Mambo who adopted it, (a) after the release, (b) two days later.

d. Acceptance of Juvenile Yellow Baboons by the Hamadryas Troop

The introduction of the yellow baboons *(Papio cynocephalus)* was the first of the whole series. The two juveniles had been captured near Mustahil on the Webi Shebeli River in Ogaden and had lived together in captivity for several months. The Mustahil baboons are savanna baboons. Males and females have the same short, yellow and gray hair, black faces and long tuftless tails which stand up from the roots. They use the trees of the gallery forest to sleep on, and their groups probably do not consist of one-male units. The yellow female baboon, Agnes, was about 2½ years old; the male, Bimbo, about 1½. After several unsuccessful attempts the yellow baboons were caught and adopted by two young adult hamadryas males.

We released the yellow baboons at the foot of the White Rock as the troop was coming in. Agnes and Bimbo began to play with each other. Several two and three-year-old males immediately approached (Fig. 47). When Bimbo began to give the play vocalization, a pseudopod of about twelve males, all juvenile and subadults, emerged from the troop. One of the three-year-olds wrestled with Bimbo and mounted Agnes. At the base of the pseudopod three young adult males appeared. Agnes and Bimbo took flight towards us and we drove them back. The three hamadryas adults threatened us whenever we chased the yellow juveniles away. Finally, we were able to drive them so far away from us that an almost adult male, whom we shall call Steady, came to stand between us and our animals. The two yellow ones now took flight into the bushes and were pursued by Steady. Bimbo climbed into the thin branches of an acacia where Steady could not reach him. Steady threatened him; at which Bimbo uttered a staccato-cough. Immediately Steady

Fig. 47. The released yellow baboon male and a juvenile hamadryas male sniff each other's mouth, a gesture of cautious approach. On the right, the yellow baboon female.

directed his threats at us. Suddenly Agnes appeared at the edge of the bushes. Steady immediately pursued her, around the sleeping rock. Finally she was caught by Steady at the edge of the rock. As soon as he held her to the ground, she stopped trying to escape. Two minutes later he began to hug her and sitting, held her close to his belly until it became dark and we lost sight of them.

Meanwhile, we had chased Bimbo down from the acacia toward the cliff, which he finally succeeded in climbing. When he began to hum, a few juvenile and sub-adult females looked down at him, but otherwise paid him no further attention. After a while, he approached a one-male unit, but the unit's females shrank back from him. Just as he was about to withdraw, however, the leader overtook him and drove him back to the unit. As darkness approached we again saw him leaving the unit.

The following morning Bimbo was seen in the arms of the young adult male Mambo. Sometimes he would disengage himself from Mambo to examine stones and grass; but never did he go more than 1½ meters away. As Mambo rose to leave, Bimbo clung first onto his leg and finally onto his belly. Both of them then disappeared into the departing troop. During the next days, as well, Bimbo never went further than two meters away from Mambo. Mambo would groom him, carry him on his back when the troop was on the move and hold him in his arms on the sleeping rock. As they were climbing over a difficult part of the sleeping rock, Mambo turned around toward Bimbo and pulled him up after him (Fig. 48). Now and again black infants would approach him and try to play with him, but he

a b

Fig. 48a, b. Mambo takes the yellow baboon male to the sleeping ledge.

Fig. 49. The cynocephalus female fails to follow the hamadryas male.

always retreated back to Mambo. After one week, he was seen allowing himself to be groomed by an adult female who came to him repeatedly despite the threats of her leader. Thus during the course of the experiment, Bimbo regressed from the status of a juvenile who took part in play-fighting with three-year olds to that of a black infant. Nine days after his introduction into the troop he disappeared.

Agnes' fate was somewhat different: On the morning after we had released her, we saw Steady as he chased her again and again over the rocks. When he reached her, she would press herself to the ground and scream, then follow him for some two meters only to sit down again (Fig. 49). Steady would then attack her again. Throughout all this Agnes remained calm and even playful, while Steady's increasing agitation became apparent through his scratching, yawning and touching of his snout. His threatening of Agnes, however, never reached neck-biting intensity.

After a while, a pseudopod of juvenile male observers was formed between the couple and the troop. Then came the crisis: As the troop was breaking camp, Steady tirelessly continued to attack Agnes and without turning his face from her began to walk back and forth between her and the troop. Agnes still would not follow him. Steady's excitement mounted and the behavior just described grew into an alarmed 'Bahu' calling. Finally the troop was lost sight of and Steady stayed alone with Agnes at the sleeping rock until the troop returned in the evening. The following morning, Steady was seen continuing his attacks on Agnes, while a pseudopod of juveniles had again formed. Twice she was chased into the swollen river below and each time swam back to shore. The second time this happened, Steady pulled her out of the water and gave her, for the first time, a severe bite on the neck. After this, she followed him closely toward the troop, but sat down again after two meters. Additional neck biting did little to improve the situation. Finally, however, she followed Steady over the cliff where they disappeared into the departing troop. In the evening Steady returned with the troop, but without Agnes. Nor did we see her again in the following days. We may assume that Agnes managed to escape enroute into the bushes where Steady was unable to find her.

e. Hypotheses

The shortcoming of our experimental attempt is the lack of a control series. By the time we succeeded in trapping any baboons at all, the field year was nearly over and we were unable to repeat the procedure with animals of the 'receiver' troop itself. Would a female trapped and released again into her own troop be retrieved by her former leader? Would an entire one-male unit be fought and disrupted by the males of its own troop because it was removed for a day? These questions will have to be answered in a more complete series of experiments. The following is based on the assumption that the procedure did not seriously change the strangers' and the troop's responses to each other.

Returning first to the question of what role familiarity among individuals plays in the social organization, it is clear that the one-male unit and the troop must be considered separately. The strange hamadryas females were immediately incorporated into units and at once took up following and grooming relations with their new leaders. Familiarity, therefore, is no prerequisite for being accepted into a *unit*, i.e., for the formation of a female-male bond.

For an entire unit, however, previous interactions or 'familiarity' with at least a part of the local troop seem to be a prerequisite for its acceptance and safety within the *troop*. The strange adult males were not chased away, but one of them was robbed of his females while the other prevented this by leading them on an immediate escape. In a similar way, during the second phase of battles, some leaders hastily led their females away from the place of fighting and, during the third phase, the bands of the troop would flee from each other. The observations leave little doubt about the function of these retreats: By fleeing from a less familiar party, the unit leaders avoid risking the loss of their females. The following hypothesis may serve as a summary. The prerequisite for the formation of larger associations in the hamadryas is not familiarity between units as such, but a correlate of familiarity: The leaders' inhibition to compete for each others' females. If this relation has not been established, or if it breaks down, the units avoid each other. Normally, the inhibition is effective throughout a troop. Only if units keep intermingling, as after artificial feeding, the inhibition between bands becomes insufficient; bands are no longer safe near each other, and the troop splits up. By *not* disintegrating under these circumstances, the band demonstrates its higher resistance against internal competition for females.

The second hypothesis to which the transplantations give support concerns the status of the hamadryas female and the character of the one-male unit. According to DeVore (1962, and personal communication), anubis males chase adult strangers away from their group, males and females alike. It is interesting to note that in the anubis society, females are free to move and to engage in interactions throughout the troop. In comparison, the hamadryas female is merely a social appendage of her male whose constant watch virtually prevents her from contacting other troop members. As a stranger, she is retrieved by a male with as little preliminaries as if she were a young juvenile. This similarity again supports the hypothesis that the hamadryas one-male unit in several respects resembles a mother-child relation. The dependent hamadryas females, like juveniles, are precluded from full troop membership, and therefore they must not qualify for it when they join the troop.

Considering the immediacy with which a male-female bond may be formed, its strength is impressive. When females were released with their leader they followed him as long as possible, one of them even leaving her infant. That unit ties take precedence over troop ties was again shown by Steady when he gave up contact with his troop for the sake of a female of another species whom he had not known a day before.

The outcome of the yellow baboon experiment is of some interest in relation to eventual transplantations across the species border. The juvenile yellow baboons were accepted into the hamadryas troop in the same way as were juvenile hamadryas. Steady's relation to Agnes was virtually a copy of an initial unit in the hamadryas style. If hamadryas males actively retrieve females and juveniles of other baboon species, a hamadryas population on the border of the species area may be open to an occasional genetic influx from its neighbors. The dark pigmentation of the hamadryas near its western border and the observation of possible hybrids (p. 23) should be considered with this in mind.

VII. THE COORDINATION OF TRAVEL

1. SPATIAL FORMATIONS OF THE ONE-MALE UNIT

We shall now turn to the coordination between the one-male units of a troop on the march. The basic pattern of coordination may be derived from the protocols themselves. The critical reader therefore should first read the protocol beginning on p. 124 and draw his own conclusions.

In the preceding sections we have become acquainted with the one-male unit at the sleeping rock and during the midday rest at the watering place. We have missed nothing of such social behavior as fighting, sexual behavior, infant care and social grooming, since all occur mostly at resting places (Table XV, p. 169). Social interactions on route are far less conspicuous: Bodily contact is absent altogether and even members of a one-male unit do not exchange more than an occasional glance or contact grunt for hours on end. Moving and eating are the main occupations. Nevertheless, social relationships find their expression in distance, formation, and direction of travel. The behavior patterns which coordinate the moving troop have been treated only summarily in many field reports though they are of vital importance for the troop. They guarantee cohesion and formation and thus play a part in the utilization of food resources and the avoidance of predators; further, they determine the route and with it, the choice of sleeping rocks, feeding and watering places. With some exaggeration one might say that in the safety of the sleeping rock, social relationships are being established, whereas in the open field they are in function; here, their ecological value comes to the test, and during this subtle functioning the purely establishing patterns of behavior are suppressed.

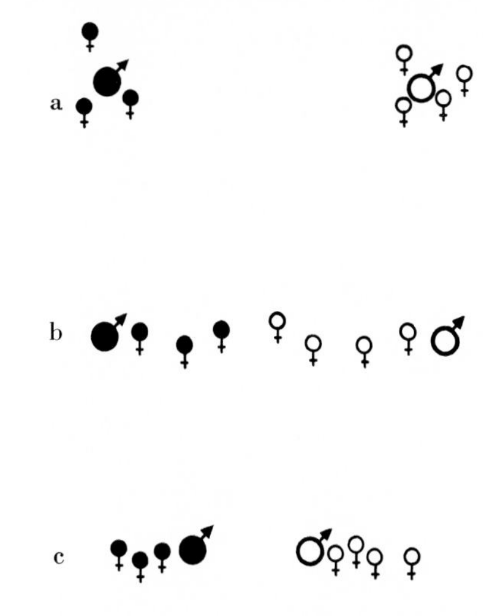

Fig. 50. Typical formations of one-male units at rest (a), during travel (b), and during aggression between neighboring unit leaders (c). Units are distinguished by black and white symbols, respectively. Infants are omitted.

The change in mood which occurs during the transition to travelling is already reflected in the spatial arrangement of the one-male unit (Fig. 50). In the resting unit (a) the females will sit more or less in a circle around their leader; attention and behavior is concentrated on the unit itself. As soon as the troop begins to move, however, the social vista broadens and the location and movements of at least the neighboring units become important. The females, instead of surrounding the leader, keep on his safest side. They either line themselves up between their leader and a previously disregarded neighboring male, or they form a line directed towards the center of the troop. Very frequently, two units will walk in a line (b) in which the males take the front and rear; or if the column is moving abreast of the direction of the march, the males walk at the wings. The differences between resting and marching moods are underscored by the change in escape distance towards the observer (see p. 170). It is possible to approach a resting troop, whether in the safety of the cliff or at a watering place, to within half the distance of that of a travelling troop.

The marching order of classes in a number of large travelling parties has been treated statistically. The frequency with which adult and subadult males appear at the front of their columns is twice that which would be expected under conditions of random mixing. On the other hand, these two classes were found bringing up the rear with a frequency equal to chance. In parties in which there are only two males the marching order is less flexible and the front and rear are almost always taken by the two leaders.

2. THE TWO-MALE TEAM AND THE WAY IT WORKS: THE I-D SYSTEM

With the changeover from the resting mood to the marching or troop mood, relationships come to light between one-male units which were latent at the sleeping rock. The smallest entities which thus come to the fore are the associations of two units coordinated by their leaders, the 'two-male teams'. These teams are among the most fascinating phenomena in the social organization of the hamadryas. The patterns of their functioning give us the key to understanding the coordination of troop movement. When travelling, the two leaders frequently interact with each other, while their females, as before, restrict their interactions to their own units. Each team consists of a younger and an older unit leader. Altogether, I was able to watch continuously eight such teams for periods of 30 minutes in some cases to 13 hours in others. The term 'team' is chosen for the following reasons:

1. The direction of the march results from a compromise between the intentions of the two leaders, not from the intentions of the dominant male alone.

2. The positions of both males are adjusted to one another in the marching order: If one of the leaders takes the front, the other will wait to take the rear.

3. Even after continued differences about which direction to take, the two leaders never threaten each other.

The way in which the two-male team operates is illustrated by the following protocol which was taken of a two-male team that lived alone at the Ravine Rock for two days. Circum (p.34), the younger of the two males, was in his early prime and led a unit consisting of four females and three infants. Pater, the elder, had only one female with a two-year-old juvenile male. The protocol has been shortened in order to avoid needless repetition. Omissions are indicated by dots. Figure 51 shows the first part of the described route which is presented fully in Figure 66. The direction in which each of the males proceeds should be noted.

07.05 h The two units are seated above the cliff which stands over the right bank of a northerly flowing river. The adults are engaged in grooming while their young play together.

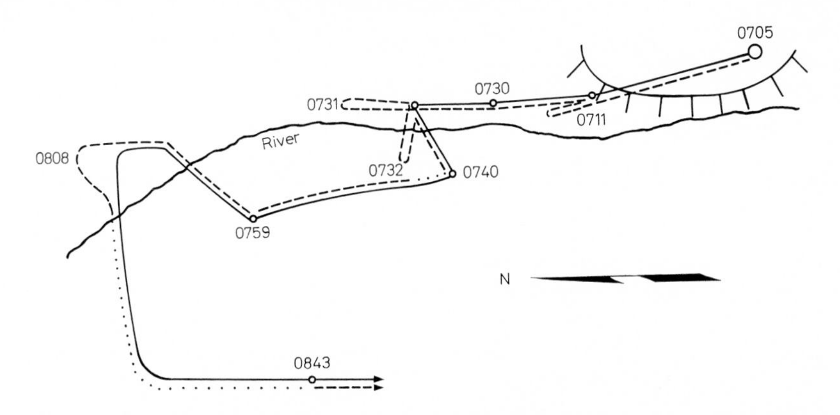

Fig. 51. First 150 meters of Circum's and Pater's travel route (see text). Small circles indicate places where the two units sat down. The solid line marks Pater's path; the broken line shows Circum's path when he walked ahead, and the dotted line when he walked behind.

07.09 h Circum rises and uttering a contact grunt goes over to Pater. Both exchange glances. Circum goes ahead to the north. All the animals follow. The marching order is as follows: Circum, his females, all the young, Pater's female, Pater. The troop proceeds to the watering hole at the northern foot of the cliff.

07.11 h The entire party seats itself in the riverbed. Grooming and play.

07.30 h Circum utters a contact grunt and goes northwards along the riverbed. Again the entire party follows for some 20 meters and then stops.

07.31 h Circum again rises; he briefly looks back at Pater and then goes another 30 meters northwards. No one follows. He stops, comes back until he is only 20 meters away from his closest female and sits down. All the while Pater has been watching him.

07.32 h Circum rises and begins to move west, straight across the riverbed. Only his youngest female follows him. After a few seconds both come halfway back and sit down.

07.33 h Circum sets out again, this time in a southwest direction. Now, Pater raises, and the whole party follows Circum in the same marching order as above.

07.40 h On the left bank, Circum again begins to point northwards. The others follow him. After 100 meters they all sit down, then they climb an acacia and begin to eat the blossoms.

07.58 h Pater climbs down and sits next to the trunk facing northwards. Circum immediately climbs down to him; Circum stops a second near Pater while the two of them turn their faces towards each other. Then Circum continues further north. Pater allows females and young to pass and then follows at the rear. 20 meters ahead, everyone stops.

08.07 h Circum rises, looks back and proceeds. All follow. After a few meters he swerves to the right, leads the troop back to the right bank and continues northward. At this, Pater slowly moves to the front of the column. 10 meters ahead, he overtakes Circum and as he reaches the front turns westward back to the left bank. The females and young swing around him to the west like a rope. Circum, on the rope's end, continues northward a few paces but then follows the rest across the river. For the first time the marching order is reversed: Pater, Pater's female, Circum's females, Circum. The young are scrambling about at the side of the column. Pater now leads the party westwards and then to the southwest. (Note that this was the direction in which he ultimately followed Circum at 07.33 h).

08.43 h The party is seated on a hill to the west of the river.

08.52 h Circum rises and leaves his place at the rear of the column. Walking to the front, he passes very close to Pater (30 cm) and now precedes the others again, going south. The females follow him. Pater waits and takes up the rear...

10.35 h A heavy downpour. Everyone is sitting in rest formation.

11.00 h The rain lets up. Pater rises, glances over at Circum and begins to move southwards. Circum follows some meters behind the rest of the column, but then turns off in a northeasterly direction. All his four females and all the troop's young follow closely. Pater and his female remain standing still, looking at the others withdrawing. Finally, they too follow and Pater brings up the rear...

13.11 h Rest. The two units are sitting 20 meters apart in well defined rest formation. Play and grooming.

13.15 h Circum rises and slowly goes to the north, turning back to the troop several times. His females follow him; some even passing him as they search for food in a northwesterly direction.

13.19 h Pater who was sitting southmost, rises and begins to move in the opposite direction, to the south. Within three seconds, all of the females and young follow him; finally, even Circum.

13.23 h After going 40 meters, searching for food, Pater sits down. His unit stops. Circum's unit continues to look for food.

13.44 h Pater and his unit walk back, northwards, some 50 meters, to Circum's unit. Whereupon Circum immediately begins to walk rapidly towards the northwest.

13.45 h Pater sits down. Despite the fact that Circum cannot see him over a rise, the whole column, with Circum at the end, swings about Pater towards the northeast and sits down.

13.46 h The two units depart and begin to move rapidly toward the east, in the direction of the Ravine Rock, Circum is leading; Pater, brings up the rear...

16.20 h The troop enters the Ravine Rock in the same order.

From these and similar observations, it became clear that lead and marching direction are changed in accordance with well marked rules. I shall refer to examples in the above protocol by the times indicated.

1. Choice of route is a matter concerning the adult males. The protocol shows that unit leaders do indeed tend in different directions.

Circum continued to turn northwards until early afternoon. The other parties of the Ravine Rock troop had taken this direction on their way to the Rotten Rock some days earlier. Pater, on the other hand, maintained a southwest route which led back to the Ravine Rock. Circum intermittently complied by leading south-westward (07.33 h, 08.52 h), but only after 13.46 h did he lead the way to the Ravine Rock without trying to swerve off to the north. Despite these differences, the units remained together.

2. Females are to some extent capable of influencing the direction of the march by immediately following one or the other of the two males. Whether or not Circum's advances were successful seemed to depend somewhat on whether or not his females followed him (07.31 h, 11.00 h, 13.19 h).

3. The younger of the two males most often takes the lead. He is followed by his females, then by the females of the older male and finally by the older male himself. The formation in which the two males walk parallel to one another at either end of the string of females occurs mostly when the males are striving in different directions. The chain of females holds the party together even when the two leaders are not in sight of each other (13.45 h). The general centripetal tendency of females might be one of the attractive forces that keeps units together.

4. Each of the males assumes a different role. Shifts are usually initiated by the younger of the two males. He also presses the party to break camp by leaving it in a given direction (07.30 h and 07.31 h). His advances, however, are only taken as 'proposals'. The older leader watches these advances from the back but only follows when the proposal concurs with his own intended direction (07.33 h). Thus, in the two-male team, initiative and decision are divided functions: The younger and less influential leader takes the *initiative* or I-role, while the *decision* making or D-role is taken by the older. At break of camp in the morning, the I-male probably finds out about the D-male's intentions by trial and error. If the D-male remains seated, then, very often, the females belonging to the I-male also remain in place (07.31 h).

5. Now and again during the course of the day, the I-male will come back to his first proposal. In this case, the D-male may come to the front and take the lead for a time (08.07 h), until the I-male is ready to continue the lead in the direction indicated by the D-male (08.52 h). Or the D-male can simply remain seated (13.45 h) or precede

the troop in his own direction (13.19 h). The zigzag pattern of the described route shows that the D-male's power of decision is not absolute.

6. At a change of direction or at departure after a rest period the leaders exchange glances or one of them goes over and presents either face or anal field to the other (07.09 h, 07.31 h, 07.58 h, 11.00 h).

3. PATTERNS OF INTERACTION BETWEEN LEADERS

With the exception of fighting, and its particular behavior patterns, those just mentioned under (6) above are the only patterns with which unit leaders make discernible contact with each other. The various forms of this class of behavior all occur in similar situations and appear to have a similar function which we have designated as 'notifying'.

In the most pronounced form of this behavior one of the males slowly approaches a seated neighbor, while both look at each others' face (Fig. 52 a). As soon as the approaching male has come very close (30 cm) to the other, he turns abruptly and presents, often with an erection, his anal field, i.e., the highly colored ano-genital region and the ischial callosities (Fig. 52 b); then he immediately begins his peculiarly hasty retreat. When the presenting male turns, the seated one will often bend forward as though to get a better view of the other's hind region, or sometimes will lift the back of his hand to his partner's scrotum (Fig. 52 c). He then gazes at the withdrawing anal field (Fig. 52 d). The hesitating approach and the hasty retreat of the presenting male is due to the fact that the usual distance kept between leaders (about 2.5 meters) is sharply reduced during the actual presentation. In a weaker variant of the behavior the approaching male comes from behind and passes the other very closely. As soon as his anal field is abreast of the other male, he stops for a second and looks at the other's face. Or the notifying male might come from the front, stop, exchange a glance with his partner and proceed without presenting his anal field at all. In the most inconspicuous form, the males simply look at each other for a second over a distance of several meters.

The constant components of notifying behavior, then, are an exchange of glances and facial presentation which is followed in the

Fig. 52 a–d. Notifying behavior among adult males. See text.

majority of cases by presentation of the anal field. All forms of this pattern are carried out only by adult males toward other adult males. An early form of this for subadult males will be described later.

Among primates, the presentation of the anal field is primarily an invitation to copulation by the female. In addition, females and other classes use this gesture to indicate submission to a superior partner or to a potential aggressor. Among adult hamadryas males, however, presentation of the anal field is not an expression of submission. An influential adult in his late prime can present to a non-influential young adult, as well as the other to him. In the next section we shall see that although the male in the I-role presents more frequently, he does not do so exclusively. In the early stages of the two-male team, the older male assumes both D and I roles; here it is he who primarily presents to his younger, subordinate partner who hardly takes part in leading the party at all.

The function of notifying can be inferred from its temporal concurrence with changes in relative location. A male who presents to a neighbor in this manner will either proceed to settle in the vicinity of this neighbor; or if he has been sitting near him, will leave him after this presentation. Before the troop breaks camp from the sleeping rock, presenting and the exchanging of glances among males increase manifold. An analysis of some 70 cases of notifying has led to the following interpretation: The exchange of glances and the presentation of face and anal field alert the neighbor that the spatial relationship between him and the active partner is about to change in a way which is important for the cohesion of their party. The behavior thus imparts to the ensuing shift a special significance. It causes the receiver to pay attention to the subsequent movements of his partner. The analogy of notifying behavior to the human greeting or the military report is obvious, though in those cases the front alone is presented. The patterns for meeting and parting are, for the hamadryas, as in many human forms of greeting, the same.

The examples and protocols in the next section will support this interpretation and illustrate a few special uses of the notifying behavior.

1. A male may notify his neighbor before he leaves him.

In a still stationary party of the troop, three males are seen as they get up within a few minutes of each other. Before leaving the party, each of the three goes to notify the same seated neighbor.

2. A male will notify a new neighbor upon settling down in his immediate vicinity. Between taking his place and notifying there is a time lapse ranging from 3 to 8 minutes. The first of the following illustrations occurred between Circum and Pater on the evening of the march described above.

Pater goes to his usual sleeping ledge. Half a minute later, Circum and his unit move in and sit two meters away from Pater. Six minutes later Circum rises and uttering a contact grunt approaches Pater. Halfway there he presents and returns quickly to his place. The same scene occurred the following evening with a lapse of eight minutes.

Before the break of camp a leader is seen moving past Rosso to sit down three meters away from him. Five minutes later, he approaches Rosso frontally, presents his face and returns to sit in his newly acquired spot.

3. The receiver may respond to the notification by moving off in the same direction as the notifying male. In this way the notifier can fetch a lingering male or push him on ahead of himself.

On route. A band has just left its resting place; only one party consisting of about 10 animals remains seated. At this, one of the departing males returns and presents to one of the seated males. Immediately the receiver, and after him the whole party, rises and follows the notifying male.

Before break of camp. A leader coming from the center of the troop sits down four meters behind the advanced periphery. Three minutes later he approaches the foremost male, presents his anal field and walks back to his new place. During the notifying the receiver rises and then advances five meters, ahead of the notifyer.

4. Older and more influential males may present to younger ones who may respond submissively.

Before break of camp. An adult male approaches another, barely adult male. The younger avoids looking at the oncomer's face. Smacking his lips, the younger rocks from side to side as though trying to get a look at the other's anal field ahead of time. The older adult presents his anal field in the usual manner and then withdraws.

In rare cases the receiver is subadult.

An adult male comes to a four-year-old. He notifies and withdraws quickly. The four-year-old follows him.

5. Adult males sometimes execute an anal field presentation into empty space without a receiving partner. The cause of this behavior is unclear. That it is not a matter of redirection away from a feared partner is evident from the fact that even influential D-males notify into space.

6. The receiver may ostentatiously avoid looking at the notifying male.

Before break of camp. The troop begins to advance gradually westward. Far at the back of the troop, three males who had previously joined in an advance to the northeast still hold out. The one farthest at the back rises, goes to the second and stops next to him for a second. As he begins to go on to join the troop, the receiver abruptly lowers his head and looks to the ground.

Figure 53 shows this sinking of the head which is strongly reminiscent of the cut-off behavior, described by Chance (1962), whereby the receiver screens himself off from certain social signals.

We can only guess at which of the presented stimuli are important in notifying. The behavior first appears in the male ontogeny among subadults in the form of facial and genital presentation. Here the actor presents his flank to the receiver, at the same time placing one leg behind him in such a way that the receiver can see and grasp

Fig. 53. The male on the right abruptly lowers his head as another male (left) notifies him.

his penis. The last trace of this grasp can be seen among adults in the perfunctory touching of the presenter's scrotum with the *back* of the receiver's hand during notifying. Which parts of the hind region are effective in notifying is, however, not important for the following hypothesis of its function: The bright red hind region of the males and the notifying pattern itself are, as far as is known for baboons, unique among the hamadryas. The anal field and the face are especially conspicuous among old, little pigmented males (Fig. 54). Furthermore, the anal field varies from male to male (at least in the eyes of the human observer) just as much as the face and is therefore suitable for identification purposes. In a highly organized society, it is not merely a question of keeping together an anonymous mass, but rather of not losing particular individuals in the crowd. In a hamadryas troop, the conspicuous adult males alone form the connecting knots in the network of animals. Above the unit level, individuals without grey mantles do not have to be taken into account by others. While sexual dimorphism thus delivers information concerning an individual's membership in the relevant *class*, notifying may allow the receiver to become acquainted with the *individuality* of the male in front of him, first by being presented with his face and then his anal field, which a few seconds later may become the only cue with which to identify him. In addition, the white-red-white signal might release a following response in the receiver.

Fig. 54. An old (left) and a young adult male. The whitest hair in hamadryas is found on the older males' body, around the face and in a belt aroung the anal field. Thus, both features presented in notifying behavior are red surfaces surrounded by white.

4. STAGES IN THE DEVELOPMENT OF THE TWO-MALE TEAM

Having arrived at a knowledge of the patterns of coordinating behavior, we can again turn to the mechanisms of troop movement. In the light of the previously described development of the one-male unit, the developmental stages of the two-male team are easily discernible. The consistent difference in ages between the two males suggests that the younger may be a former follower of the elder. Figure 55 shows the composition and typical marching order of observed two-male teams in order of the presumed life cycle of such a team.

In the section on followers, it was seen that subadult followers do not take a particular position in the moving one-male unit and can as readily be found at the end as at the side of the column, or in the middle of the unit's females (Fig. 55 a). While in the ensuing stages, team partners never separate during a march, subadult followers will often leave their unit. The direction of the march is initiated and decided upon solely by the unit leader.

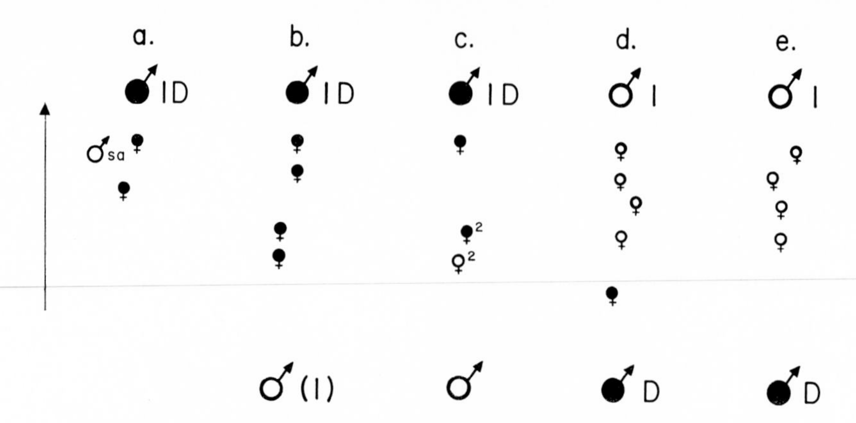

Fig. 55. Composition and marching order of two-male teams in the order of a team's presumed life cycle. Black symbols represent the older male's unit, white symbols the younger male's unit. sa = subadult, 2 = two-year-old; I = initiator, D = decider. Individuals whose ages are not indicated are adult. Juveniles and infants have been omitted. The marching direction is indicated by the arrow.

As soon as the one-male unit contains an *adult* follower, the rest of the unit responds to his position and his marching direction (Fig. 55b). The females are now strung between the two males; though they keep closer to the leader, who walks at the head of the column. If the follower happens to take the initiative and, in consequence, deviates from the direction of the march, the leader will sometimes approach him and present to him. In these cases the younger male usually complies. The leader notifies the follower more often than the follower him, and still assumes both I- and D-roles. Two examples of this type of two-male team were observed. One of them on route is described on p. 147.

The next stage (Fig. 55c) was observed only once. In this team the younger male already had a two-year old female of his own (initial unit). The two units had spent the night alone on the Ravine Rock and were sitting above the cliff in travel formation with the older, red-faced male at the head. (Again, the following protocol contains only the behavior pertinent to the break of camp itself).

07.40 h The column is 12 meters long. The younger male moves to the front and presents to the elder, who immediately rises in response and advances some 10 meters in the direction of the future march. His females follow and sit around him. The younger sits where he did the presenting.

07.47 h The older male scratches himself and yawns repeatedly (behavior related to slight conflict, p. 144).

07.50 h The older male rises and looks in the direction of the intended march for some four seconds; he scratches himself and touches his hand to his mouth. Now he approaches the younger male and presents to him, then begins to move in the marching direction. The entire party follows, with the younger male coming at the rear only after a fifteen-second delay. Everyone sits down again.

08.09 h The older male rises again, uttering two contact grunts, and begins to move on. His females follow immediately; the younger male and his female only after seven seconds. Again the whole party sits down.

08.10 h The younger male moves toward the front. As he reaches the side of the older, the latter rises and walks alongside of him. Everyone sits down; the younger in front of the older for the first time. The younger male scratches himself repeatedly, yawning and touching his mouth.

08.25 h The older male moves past the younger uttering a contact grunt. The whole party follows with the exception of the younger male who remains seated.

08.27 h The younger male moves up.

08.28 h The two units advance twenty meters with the old male in front and the younger male in the rear.

08.30 h Final departure in the indicated order.

Although the younger male twice provoked the older male to proceed, he himself never indicated a marching direction and almost always stayed at the rear of the column. Initiative and decision still lay with the older leader.

In the next phase of the two-male team (Fig. 55d), which has already been described in the case of Pater and Circum, both males are intensively concerned with determining the direction of the march. I- and D-roles are separated. Four examples of this type of team were observed travelling. The D-males had 1 to 3 females, the I-males 3 to 4. In two cases the I-male's obtrusive presenting and frequent goings back and forth was in sharp contrast to the D-male's impassivity and rare changes of location. The following protocol which begins 80 minutes before the White Rock troop broke camp shows this contrast.

07.02 h Smoke, a male in early prime, is sitting with his three females at the head of one of the troop's pseudopods which is extending westward. Fifth from the front, toward the center, Silver's unit is sitting. Silver seems to be older than Smoke.

07.13 h The rear of the pseudopod begins to draw back with only the first six units remaining seated.

07.19 h Smoke now too begins to lead his females back, notifying Silver in passing. Silver remains seated and now forms the head of the pseudopod.

07.20 h Smoke leads his females back to the front again and sits down next to Silver.

07.21 h Smoke rises, presents to Silver and again goes back toward the center of the troop; this time Silver follows.

07.24 h Smoke moves to the front again, ahead of all the others. Silver remains.

07.38 h Smoke returns, sits close to Silver and looks around at his females who immediately run to him screaming. Smoke takes them on a following procession.

07.57 h Silver shifts a few paces towards the center. A minute later Smoke follows him and insistently presents to him for several seconds; then he sits down, surrounded by his females. Silver draws further back, closely followed by Smoke and his females. Silver's own female is still sitting at the tip of the pseudopod 30 meters away, with her back to Silver. Silver looks over to her and yawns.

08.00 h Smoke sits down immediately in front of Silver and looks him in the face. He then goes a couple of meters to the side. Silver does not respond.

08.02 h Smoke again steps up to Silver, presents his face and anal field and sits down in his former place.

08.04 h Silver slowly goes to his female at the head of the pseudopod. While he is still eight meters away from her, she scratches herself, but does not get up. Silver now sits down five meters away from her, looks at her and yawns. She does not respond. Smoke and his females come up and sit down close to Silver.

08.05 h Silver goes over to his female and sits down with her; she grooms him. Both units are now sitting behind the head of the pseudopod.

08.13 h A large number of animals are coming from the center of the troop, led by a red-faced male. With the exception of Silver and Smoke, all units in the pseudopod rise and precede the approaching troop center.

08.14 h Smoke and his females rise and also follow the passing column. After 10 meters, Smoke stops and looks back; he returns to Silver.

08.16 h Silver rises and follows the column. Smoke and his females follow him. The troop departs.

As in Pater's and Circum's team, Silver, the D-male, controls his female far less strictly than the I-male. That Silver finally goes over to his female over a distance of 30 meters and thereby draws Smoke's unit with him, shows how females in D-units can influence the movement of a party.

The aged male will finally release his last female from his control; but even then he can still remain a member of the two-male team. Figure 55e shows the most frequent formations taken by the only two-male team of this type which I was able to follow for a sufficient length of time (80 minutes).

12.30 h I accidentally come within 12 meters of a unit. The male springs into the air and then runs 20 meters away where he sits down. All four females run after him and sit down close to him. 15 meters away from them an old male sits down alone...

13.14 h Both males move parallel to each other with the unit leader somewhat ahead and the females in a line between him and the old male.

13.16 h The unit leader sits down. All four females come up to him. The old male overtakes them and sits down 15 meters in front of the leader, alone in the shade.

13.19 h The leader rises; this is immediately responded to by scratching on the part of the old male. The leader approaches the old male with his females. 30 centimeters in front of him, he turns to the right with a rapid presentation. By doing this, the foremost female comes directly in front of the old male. She recoils and runs to the leader who is still moving towards the right. The females follow him but the old male remains seated.

13.24 h After a short rest. The old male moves off in a new direction from his place in the rear. After half a minute the unit leader and his females swing around and follow him, somewhat off to the side. After 20 meters, the old male stops, allows the younger male and the females to overtake him and takes the rear . . .

Although the party's females clearly belong to the younger male, the old team partner still displayed the traits of a D-male: He would bring up the rear and without further ado determine a new marching direction which the younger partner accepted.

Because of the difficulty of finding and following particular units in a moving troop, I know nothing about the stability of a two-male team. Only in one instance did I come upon an identified two-male team on two separate occasions, but this with an interval of only 13 days. Whether or not the males of such a team undergo all the stages in a team's development together is also not known. More important than the permanence of such a team, however, is the fact that in every team a younger male accompanies an older and that the older plays the decisive role in determining the direction of travel even if he leads fewer females than his younger partner. Thus, access to females and influence on troop movement are not combined in an all pervading status of dominance as in other baboons, but tend to be separate functions of two male age classes.

The age difference between the team mates makes the two-male team an especially suitable social entity for handing down non-genetic information from one generation to another. The position of follower may acquaint the subadult male with the behavior of a unit leader towards his females, and in the later stages, the younger leader is confronted by his older partner with the behavior that coordinates the movement of units. Both relationships might further serve the tradition of travelling habits, especially in areas where the troops are extremely small.

5. THE I-D SYSTEM IN THE TROOP

The difficulty of analyzing troop movements lies in tracing the starting and stopping, and the choice and change of direction of 100 or so animals back to subtle local happenings within the troop. Only by analyzing aerial photographs could this be done completely with precision. The ecological and sociological interest in these processes, however, justifies an attempt at their analysis with less satisfactory methods.

The systematic portion of this study was limited from the very beginning to observations during the break of camp at the sleeping rock. This was because only there were we able to oversee at least one third or more of the entire troop. Our study was divided into two parts. First we observed decamping 26 times at various sleeping rocks, noting the overall changes in the troop's contours and attempting to identify the local causes for such changes. Five of these decampings were filmed in section (one shot per second) and analyzed. At 13 additional decampings, six neighboring adult males were chosen each time and their behavior recorded from the time of their arrival in the open area above the cliff to the actual departure. Whenever other males made contact with those chosen for observation, they were included, as far as possible, in the rest of the observation. At each rock only one or two directions of departure were ever strongly favored and these could often be predicted by us. It was therefore frequently possible to select a sample which would later form the head of the departing troop. Other samples of six landed in the middle of the column, while one such sample ended at the rear.

Almost all decampings take the following pattern: At about sunrise the animals move unit by unit into the area above the wall of the cliff. Here, in the sun, they devote their time to the social behavior described in the sections on the one-male unit. In the one to three hours before the actual departure, the troop's contours begin to change, at first gradually, then with increasing speed. Amoeba-like pseudopods are stretched out and withdrawn again. Such a pseudopod or runner takes about ten to thirty minutes to form. It starts when a male leaves the troop proper and sits down some three to ten meters beyond the troop's periphery. In a short time other males follow suit, sitting down 2 to 5 meters next to him. All these males have their backs to the troop. In this way a male front is formed. The area behind this front is taken up by their females and other one-male units. Usually the units shift as clear entities; when they halt, their females sit behind their males centerwards. The pseudopod is further developed by units from the center moving ahead of the foremost males. In this way the males of an advancing pseudopod front are often exchanged. Usually one or two minutes elapse before another unit in the pseudopod shifts.

In the simplest case an actual departure will occur when unit after unit at the center of the troop walk toward the pseudopod. If enough animals in the center are walking at the same time, the pseudopod's front too rises, goes ahead, and the troop departs. Frequently, however, a pseudopod will become stationary and no longer gets reinforcements from the rear. In the meantime another pseudopod may arise, independently of the first, in another direction. Often, the first pseudopod will withdraw while the second is being formed; the males at the back of the initial pseudopod going back first, followed by those at the front. A troop will form two, rarely three, such pseudopods which independently of each other advance and withdraw until the center flows into one of them, thus provoking the actual departure in this direction. The number of animals in a pseudopod is not the decisive factor; even very large pseudopods will occasionally draw back though some males may remain stubbornly in their advanced positions (p. 131).

Twenty-two of the twenty-six decampings surveyed followed this pattern. The other four showed an especially informative variant of the pattern in that a single individual released the departure: While the usual pseudopods are advancing and retreating from the troop, an old red-faced male will stand up in the center of the troop. Instead of advancing just a few meters as usual, he moves with a peculiar rapid swinging-gait towards the troop's periphery, even in a direction in which no existing pseudopod is pointing. His neighbors respond immediately with contact grunts and accompany him in rapidly increasing numbers. Many of the males at the fronts are seen looking at him. Within a few seconds the entire troop will move out in the indicated direction whereby the starter never arrives at the lead. The pseudopods join the departure in a parallel direction from their places. The swinging-gait of the starter consists of long energetic steps, with the tail slightly raised and the pelvis swinging from side to side in rhythm to his walking. On this starting march the red-faced male neither pauses nor looks back nor presents to any of his neighbors. The swinging of the hind region may be a 'notification to everyone'.

In one of the four departures marked by such a starting march, a single pseudopod had formed in the direction most usually taken by the troop. On just this particular morning, however, the usually dry river bed at the foot of the cliff was flooded and the route was inaccessible. The starter now directed a departure toward the southeast, a rarely taken but unobstructed direction (see protocol on p. 136). The protocol on p. 146 gives an additional example of a starting march.

Although all four starting marches were observed at the White Rock, it is certain that the starter was not always the same. One of the four old males was the previously identified 'Rosso', while the other three were unknown.

The similarity these proceedings bare to the I-D system of the two-male team is obvious. The males at the tip of the pseudopod who advance in a given direction but who do not release a departure are apparently I-males; so too are the single young adults who sometimes walk 20 to 40 meters ahead of the departing troop by themselves. The final flow of the troop center into a pseudopod is presumably induced by a D-male at the center. In the four observed starting marches the D-male stood out clearly: We never saw a front of males being followed by so many animals in so few seconds as were these single males during their starting marches. The relative positions of I-and-D-males in the large troop correspond with those in the two-male team, except for the D-male who is here found in the center instead of the rear. Thus Rosso, who later appeared several times in the role of D-male, was never found in a pseudopod, but was always seen in a central section of the troop, while Smoke, an I-male familiar to us, was often found near the head of a pseudopod.

6. TYPES OF COHESION AND THE EXTENT OF THEIR INFLUENCE

A closer examination of our records shows that many males play no part in determining the direction of a particular departure; neither do they form part of a pseudopod nor do they take a D-role. Those sectors of the troop which are between pseudopods, especially those which are still near the walls of the cliff remain almost stationary. Although individual shifts with their corresponding notifying behavior do occur, definite trends in the directions of moving and facing are missing. Shifts of position seem to be primarily due to affinities between neighbors. Generally, rear sections of the troop appear not to pay attention to the happenings at the front, if only for the reason that even in the most open areas above the cliff the front is hidden by trees or hills.

In the close study of six selected males and their behavior during break of camp none of the adult males notified (or otherwise interacted with) more than five other males. Only about 10 neighboring males would recognizably respond to each other's facing and shifting. The clearest communicating entities in this respect were the males of a particular pseudopod from its root to its tip. It is not known whether the membership in these entities is stable. Their size, however, suggests the possibility that they are identical to the bands discussed earlier. Those males of a troop who respect each others' females and who stay together during large-scale fights might actually be those who are on notifying terms.

As soon as one such section is ready for departure it moves off. At this all other sections get up and depart along with it, even when they have just been engaged in building up their own pseudopod in another direction. This releasing of the general departure by the first section is not integrated through interindividual contacts but rather an effect of a strong and general social facilitation. In very large troops, even this more anonymous coherence no longer causes a common departure of the troop. Troops of 400 split up into a few clearly separated columns right after departure, but still leave the rock together and in the same direction. In the troop of 750, however, who at night covered a cliff 200 meters in length, different columns departed at different times and in different directions.

Within the troop, then, two types of coherence seem to be at work: The individual relations among the males of the section (or band) integrate the movements of up to about 50 animals. Above this level, a phylogenetically older and possibly 'anonymous' mass attraction can synchronize shifts of up to about 400 animals.

7. MALE BEHAVIOR DURING DECAMPING

In the last two sections we have seen the patterns governing a troop's decamping from the sleeping rock and the extent to which its amoeboid movements are explicable by the I-D system of the two-male team. In the following, we are again concerned with the behavior of individuals. Notifying behavior has already been discussed in connection with the two-male team system and the data of the present observations have already been included in that section. Conclusions as to the function of notifying hold for the entire troop. The following illustration will call this function to mind.

Before break of camp. A barely adult male is seen coming from the center of the troop and going back to the hindmost party which is still sitting close to the cliff. He approaches an older male who is sitting there. When he has reached him, he looks into the old male's face, who, in turn, rises and both males present their hind quarters to each other. Now they both go to the center of the troop together. The entire rear party consisting of 30 animals follow.

In each of our samples of six neighboring males the direction toward which the animals were sitting was always recorded. We have termed this the 'facing direction'. After seeing the meaning behind the exchange of glances between neighboring males it is not surprising to find that the facing direction is not accidental either. Figure 56 shows the changes in the facing directions during the retreat of a pseudopod at the White Rock. A female in oestrus, belonging to male 9, instigated the withdrawal of the pseudopod at 07.52 h. Later a pseudopod was again formed in the same direction which resulted in a decamping in its direction.

The significance of the facing direction may be summarized in two points: A male faces that direction in which, given sufficient motivation, he will eventually move. Or, he will face that neighbor whom he would most likely follow. The figure illustrates the extent to which a male is influenced by the position shifts of his neighbors. Males in pseudopods always shift only a few meters at a time. None will sit far ahead of the front; some even turn back at once without sitting down. If one male returns, his neighbors will soon follow suit or at least turn around to face the center. Every shift increases the probability that a neighbor will shift in the same direction. The starting march of a D-male who struts through an entire section of a resting troop with his unit differs greatly from the usual hesitating shifts.

Males, thus are not only influenced by their own intentions in determining upon a direction, but are pulled in various other directions, e.g., by the centripetal tendency of their females and the behavior of a number of their male neighbors. Furthermore, they

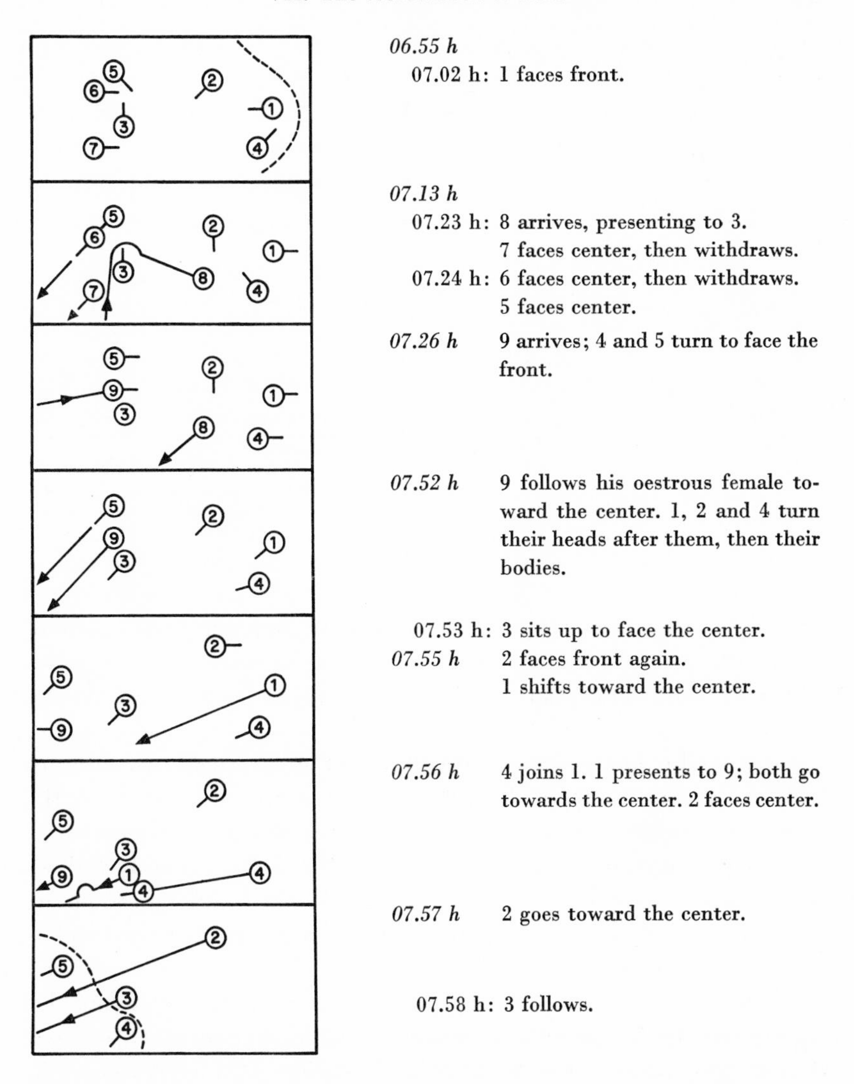

Fig. 56. Facing directions (short lines extending from circles) and shifts (arrows) of males 1 to 9 during the withdrawal of a pseudopod. A broken line represents the tip of the pseudopod. Figures represent the situation at the times printed in italics.

must respect certain minimum and maximum distances. This sometimes conflicting interplay of many tendencies is expressed in a particular behavior which may be quantitatively recorded, namely scratching. Figure 57 shows the mean frequency of this behavior in the last six minutes before and the six minutes immediately after

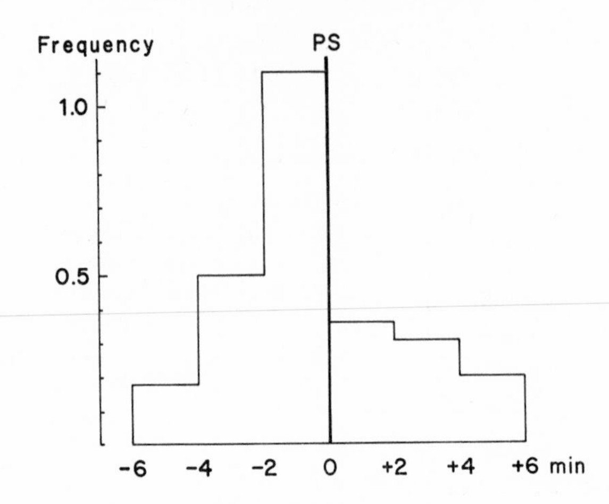

Fig. 57. Mean frequency of scratching with hands per two minute period, before and after independent shifts of position (PS) of adult males during decamping. Each of 22 males was observed during one shift.

a male shifts his position. The number of uninterrupted bouts of scratching were counted per two-minute interval. The diagram contains only 'independent' shifts, i.e., those not preceded by neighboring shifts for two minutes. As is seen in the diagram, scratching by hand is especially frequent in the last two minutes before a male gets up to shift. Most males scratch more often in the 6 minutes before a shift than in the 6 minutes after it (sign test, $P < 0.05$). In 'interdependent' shifts, where a male directly follows his neighbor or one of his females, no significant differences in the frequency of scratching by hand were found. Scratching by foot is affected in neither case.

Preening, especially among birds, has often been found by ethologists to occur 'out of context' under the influence of conflicting motivations (e.g. VAN IERSEL and BOL, 1958). Scratching in the present context parallels these classical examples of displacement activity. During decamping, unit leaders probably are forced to integrate more motivations than at any other time during the day.

8. COORDINATION ON ROUTE

The avalanche-like facilitation process at the moment of actual departure causes the troop to melt into an entity for a few minutes during the day. A strict leadership is not evident even at this time.

Young adult males at the tip are quickly overtaken and after a few hundred meters individual sections of the troop begin to move at different speeds, travelling through the savanna in parallel paths. In this first phase open river beds are frequently used. The more concentrated the troop, the faster it moves. There are no definite trails.

After an hour and a half, at the latest, parties and one-male units of the troop have scattered over an area which we can only estimate to be about 1 square kilometer. Acustic communication, however, still continues after the parties have lost sight of each other. In areas where the brush was heavy the contact grunt could be heard alternating between neighboring parties. We were still able to hear the grunt at a distance of 100 meters. The 'bahu' bark of alarm is sometimes given by visually isolated parties. In one case, upon hearing the bark, a party changed its direction by ninety degrees and joined the barking party after a walk of 600 meters.

On many occasions columns which had been separated swerved from their routes to lead us to far off parties. Such reunions have been regularly observed at watering places during the dry season. After a reunion, troops will often go through a decamping procedure similar to that of the morning.

Towards noon the observer will usually find himself alone with a small, slowly moving party or two-male team. In larger parties which are more spread out, a travelling system other than the I-D system comes into play which we may designate the 'relay-system'. Here a male will move ahead when the unit behind him comes closer. The approaching leader will then sit down exactly on the vacated spot. These points often form a chain of slightly raised seats: Rocks or low, bare branches which facilitate visual contact between males.

A unit leader is seen sitting on a raised root in the shade; his females are close by on the ground. A second male approaches from behind, notifies the first and returns to his place about 15 m away. A minute later the first male leaves the root, exchanges a glance with the other and proceeds ahead. After about another minute, the second takes his place on the root with his unit around him.

This picture of hamadryas males sitting on prominent places so frequently seen can lead one to the conclusion that these males are sitting watch. However, since they are turned *towards* the troop and are often sitting only a few meters short of a ridge that would allow a view of the next valley, this interpretation is hardly appli-

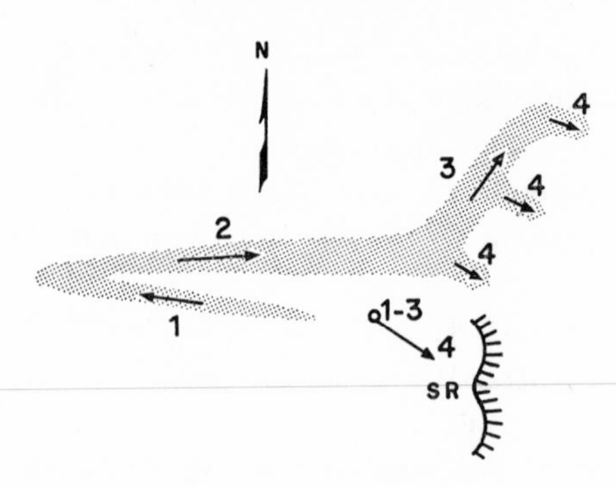

Fig. 58. The decamping of Rosso and the White Rock troop on July 26. SR = Sleeping Rock. Dotted band = The troop's paths; solid line = Rosso's path. The same figures indicate the respective location of the troop and of Rosso in the same phase of decamping.

cable. In going over a ridge one often unintentionally bursts in upon a party resting just behind it.

On the whole, there seems to be no integrated leadership of the travelling troop. That the visual obstacles of the bush present a problem not only for the observer but for the animals as well will be illustrated in the closing protocols. Here a comparison is made between the influence of a D-male, Rosso, on his neighbors at the sleeping rock and in the field. All data not directly concerned with troop movement are omitted.

July 26. On the sleeping rock. The White Rock troop has sent out a long pseudopod towards the west (Fig. 58, phase 1). Rosso is sitting at its base. At 07.45 h the pseudopod begins to flow back (phase 2) and at 08.05 h sections of the first pseudopod begin to form a second one to the northeast (phase 3). The party around Rosso has remained stationary until now. At 08.14 h Rosso rises and goes off towards the southeast in a starting march, followed by the party surrounding him (phase 4). Within 15 seconds the parties in the northeastern pseudopod rise and move out in directions parallel to Rosso's path.

August 22. On route. At midday a scattered band of the White Rock troop is seen resting at the bank of a river, northwest of the sleeping rock. Among them is Rosso with his two females and two followers.

12.27 h Thirty meters from Rosso a male is seen coming out of the brush. He sits down and looks over toward Rosso, at the same time rocking from side to side as if trying to see better.

12.35 h Rosso goes to the water and takes a drink. Halfway back he looks around toward the bushes on the opposite river bank. Another male immediately appears there and looks at Rosso.

Parties are leaving the watering place, moving towards the *south* (Fig. 59,

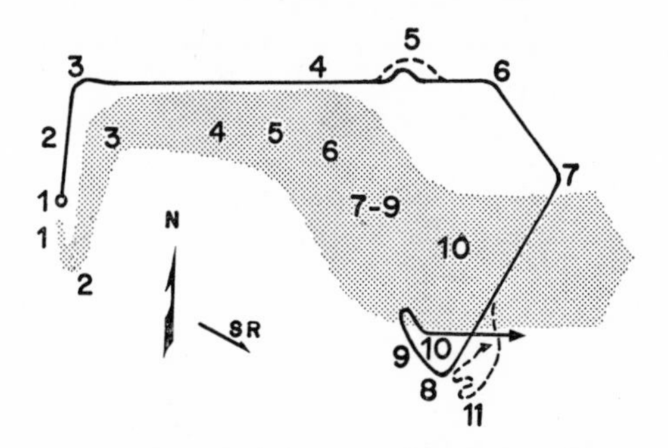

Fig. 59. The travel routes of Rosso's unit and a section of his troop on August 22, 1961. Broken line = Wave's deviations from the unit's path. The other symbols are those used in Fig. 58.

phase 1); the four one-male units next to Rosso are following them, one after the other. Rosso sits down at his former place and begins to groom his adult follower, Wave.

13.15 h Rosso departs to the *north* with his unit (phase 2). Approximately 40 animals return from the south, follow him and even overtake him.

13.33 h The troop's majority turns off towards the east; at the same time, spreading far out. Rosso follows along on the left, northern flank (phase 3).

13.48 h The same formation. Easterly direction. Rosso's unit is moving more rapidly than the troop's majority and pulls ahead of them (phase 4). I follow Rosso.

13.52 h Rosso looks back at the majority. His adult follower, Wave, turns off to the *north*. Rosso follows him for a short distance, but soon returns to an easterly course (phase 5). Wave capitulates and follows at the rear.

14.00 h Bahu barks are heard among the troop's majority which is now out of sight, some 150 meters back. Rosso looks back and turns right, in front of them. Wave takes the lead (phase 6).

14.04 h Rosso turns more and more towards the right (phase 7), so that he is now moving towards the troop's majority in a southwesterly direction. The unit swings about him with Wave at the back, so that Wave brings up the rear.

14.10 h The majority of the troop appears to be resting (phases 7, 8, 9). Rosso leads his unit in a southwesterly direction 150 meters past and in front of them and also rests (phase 8). Wave is sitting 100 meters from the unit on the side away from the troop's majority.

14.22 h Rosso goes a few meters towards the majority which is still hidden by the brush. He climbs a tree and stares towards them for a number of minutes. Now he climbs down and leads his unit, uttering a series of contact grunts, towards the troop's majority (phase 9). Wave remains behind and is lost sight of.

14.28 h The troop's majority approaches from the west (phase 10). Rosso comes alongside and moves along with it towards the east on its right flank. I go back to Wave.

14.35 h Wave is eating alone where he has been left. Soon he finds himself 200 meters behind Rosso's departing unit. Rosso moves out of sight and the last of the animals in the troop's majority move past. Wave looks about and moves a few paces towards the east. Now he begins to call 'Bahu' and starts running back and forth (phase 11) despite the fact that the majority is visibly moving past him. Finally he runs along the moving column towards the front. I follow him at a distance.

14.44 h Rosso appears on the other side of a hollow. Wave apparently sees him at the same time I do. Again he calls 'Bahu' and rushes over to Rosso. I lose sight of the unit.

Both protocols illustrate Rosso's great influence in determining direction at the break of camp. But, as soon as his section of the troop began to travel, his influence weakened. The troop's majority no longer followed Rosso's unit, rather Rosso now sought attachment to the majority. But even here the D-male played the decisive role: Rosso alone sought and found the majority, whereas the younger, though adult, Wave never recognizably responded to the movements of the troop's majority and even managed to lose his own team partner.

9. SUMMARY II: THE HIGHER UNITS

The second part of the study focussed on the relationship between the one-male units and on the integration of those units into two-male teams, bands, and troops. Among various methods of describing and analyzing the larger associations, studies on spatial arrangements and transplantation experiments were found especially useful.

The most obvious unit within the hamadryas population is the *troop*, i.e. that aggregation of one-male units and single males that spends a night on the same sleeping rock. Troops are large (up to 750 individuals) in areas where rocks are scarce, and small (as few as 12 individuals) where rocks are abundant. In the area of our broad sample the average troop size was 121. The number of animals that settle on a particular rock varies from night to night; however, certain individuals and units are nearly always present. The troops, therefore, are not stable social units in themselves; rather, troops are aggregations of smaller units which assemble on the rocks of an area in varying combinations. Their formation is determined by ecological rather than social factors.

The foraging parties one encounters during the day are much smaller than the troops one finds at the rocks. Several of these parties, which number from 30 to 90 animals, were found to be of stable size. In the evenings they were seen to congregate with other parties on the sleeping rocks. We called such parties *bands* and concluded that they were stable components of the troops.

Not all bands of an area tolerate each other on the same sleeping rock. If a band finds a rock occupied by certain other bands it may withdraw to another rock. On one occasion, when a foreign band actually settled down with the troop of our close study, there followed a general fight among males over females. Surprisingly, the local troop did not fight the intruders as a unit, but soon disintegrated into bands. We found that even regular troops split up into bands that fight or avoid each other, if one induces the one-male units to intermingle. In a pilot study on the effects of social transplantation, two one-male units were released into unfamiliar troops. One unit escaped from the troop; the other was attacked and its females were taken by the troop's males. The observations suggested that:

(1) The major prerequisite for the association of several one-male units is an inhibition in the males that restricts them from constantly competing for each others' females;

(2) The inhibition is strong within the band, moderate between some but not all bands of an area, and absent among complete strangers;

(3) The inhibition operates only as long as the units avoid spatial intermingling;

(4) Units, bands, or troops avoid each other when a low degree of inhibition between them endangers the integrity of their one-male units;

(5) The inhibition is an effect of previous interactions among the males. This was supported by the finding that the individual male in the troop interacts only with a limited set of other males, who presumably are the members of his band.

Common travel by several one-male units is coordinated by the unit leaders. In this situation it becomes apparent that two adult males of different age often have an especially close relationship and lead their units together. The development of such *two-male teams* can be reconstructed from the observable stages. In the fully developed team the younger leader initiates movements in a certain direction by walking ahead (I-role). The older male, who usually brings up the rear of the two units, accepts or rejects the indicated direction by following or by remaining seated (D-role). The leaders 'notify' each other by a set of presenting gestures when they change their relative locations.

The band determines the direction of its departure from the rock by using the same division of roles. I-males slowly advance from the troop in one or more directions, but the actual departure is released by D-males walking from the center toward the periphery in the direction of their choice. Notifying behavior is exchanged only within band-size sections of the troop. Once a band departs, the other bands interrupt the deciding procedure and follow the departing band.

10. INTERPRETATIONS AND COMPARISONS

A balanced description of all components and aspects of a complex society is a task beyond the capacity of a single team of investigators. The best one can do is to make one's biases as explicit as possible and thereby show what has been neglected. In pointing out what appear to me the constituent elements of the hamadryas society, I also wish to reveal the gaps in our picture of this society.

There is one thing to be said for our biases: If they were imported from Switzerland, they were not imported as conscious hypotheses. We sincerely wished to describe the hamadryas society as we found it, rather than to look specifically for a system of male superiority. A general, preliminary field study is an opportunity to widen concepts and to discover unexpected phenomena. To organize it around concepts imported from laboratories, zoos, and even from other species is to limit its potential. Accordingly, we tried to hold our starting set of explicit ideas down to the general biological concepts contained in our list of questions (p. 4).

During the field year, our attention became gradually focussed on the extreme differentiation of male and female roles. Essentially, the hamadryas society is an exclusive society of males, the females being distributed among its members as controlled dependents who have no active part in the life of the higher units (Fig. 60).

Consequently, the hamadryas organization is based on two types of social bonds: The bond between the adult male and his female, and the bond among the adult males. Both types develop in an ordered ontogenetical sequence, whose early stages may induce behavioral modifications required in the mature stage.

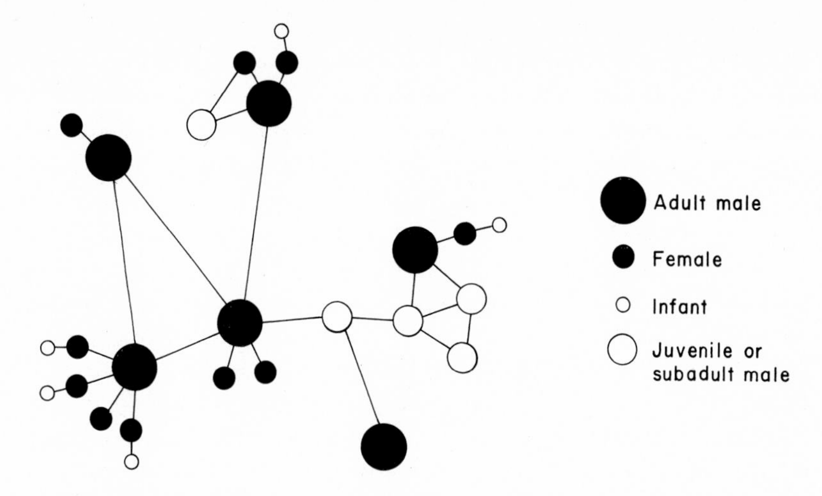

Fig. 60. Social relations in a small, hypothetical hamadryas band, showing the double role of adult males as unit leaders and band members.

The term 'bond' is used here to designate a relationship characterized (a) by mutual attraction and (b) by the specific exclusion of non-members who, by their sex and age, are potential members. Both types of bonds among adult hamadryas baboons share these basic qualities, but they differ from each other in several other respects:

(1) The male-female bond is aggressively enforced by the male, and it is therefore doubtful whether the female would spontaneously be as constant a follower and as exclusive in her interactions with adult males as she is forced to be in the actual social system. In contrast, male-male bonds are apparently never enforced by threats, and the partners may temporarily leave each other.

(2) The male-female bond is a bond between two animals and has little implication for the troop as a whole, whereas males establish multiple bonds among each other, and these are what holds a band together. Again, the exclusiveness of the female may be an effect of her male's intolerance, although the nearest-neighbor method has shown that females prefer dyadic relationships even before they fall under the male's control.

(3) The transplantation experiments have shown that strange males are not easily accepted by other males. In contrast, a male-female bond can be established within seconds by a simple transfer of the partner's role to a stranger.

The possessive, enforced, and easily transferable bond between the hamadryas male and female seems to be atypical of primates, but even as a unique case it is of general significance. It shows that a stable, exclusive bond between male and female, which sexual attraction alone fails to create, can be established by non-sexual motivations, as a kind of foster parent-child relationship.

No evidence of bonds among adult females was found in our study. The centripetal tendency of females to remain in a large aggregation appears to be based on a general, 'anonymous' attraction among conspecifics, not on specific bonds. If the female must choose between following her male and following the band, her bond with the male proves to be stronger than the attraction of the band.

Thus, the immediate cause of the existence of bands and two-male teams is the attraction among their males. At all ages before and after they have females, subadult and adult males preferentially sit close to each other and groom. At the unit-leading age the male-male attraction is less obvious. At close distances, it is now counteracted by the males' tendency to keep their units from mixing. The leaders avoid each other's proximity and never groom each other. When notifying, they penetrate another leader's unit space hesitantly and quickly withdraw again. Most of the time, the unit leaders are spatially isolated from each other in the centers of their social territories and merely communicate at a distance. Nevertheless, their mutual attraction is strong enough to outweigh the avoidance released by the proximity of their harems.

It is not only in hamadryas that the presence of females counteracts the attraction between the males. The males of the Hanuman langurs of Dharwar, India (SUGIYAMA, 1966) and of the patas monkeys in Uganda (HALL, 1965) are sufficiently attracted to each other to form purely male groups, but in contrast to the hamadryas, they do not tolerate each other in the presence of females; in their heterosexual groups only one adult male is found at a time. The groups of males and the heterosexual one-male groups all live apart from each other in these populations, whereas in hamadryas, they all unite in a higher unit, with no isolated males or male groups.

The extreme degree to which the hamadryas society is patterned by males may be emphasized by a few comparative statements which demonstrate the relatively independent status of females and the existence of strong bonds among them in other species. Hanuman langur and patas females remain together and travel normally when the male temporarily leaves them; patas females direct the group's travel even when their male is with them; in the course of SUGIYAMA's (1966) field experiment, two groups of females retained their identity and travelled independently although their common male leader attempted to drive them together. Female anubis baboons are relatively free to interact with all adults of their groups. Strange female anubis baboons were driven away by the males of a group which they tried to join (DEVORE, personal communication). In contrast, leaderless hamadryas females were never seen as detached subgroups; they were easily herded by their males; and as strangers, they were at once accepted and distributed among local males.

In characterizing the nature of the multiple bond of the hamadryas males, the concept of a dominance order is of little use. The males can be ordered according to their number of females, or according to their influence on travel directions, but the two orders do not coincide. Dominance in the usual restricted meaning, as the ability of one animal to displace another from an incentive, can be assumed to exist among hamadryas males, but its manifestations in troop life are minimal if compared with anubis baboons or rhesus macaques. The important functional order among the males of a hamadryas band is based not on the ability to displace, but on the amount of attention and compliance received from other males in the choice of travel routes. Since old males rank highest in this attention order, it cannot be based on fighting ability. The obvious aspects of this functioning order are appropriately described as an 'attention structure', a term introduced by CHANCE (personal communication). The only non-aggressive set of gestures exchanged between unit leaders, 'notifying', serves to attract a neighbor's attention to the notifyer's impending movements. The sexual dimorphism as to color may well support the functioning of this attention structure by conspicuously marking its members in the crowd of females and young. Significantly, the old males, who rank highest in this order, are most conspicuously light-haired.

Insofar as a social organization is determined by the species' genetic potential, it must have an evolutionary history which can, in the absence of fossil records, be inferred only from the organizations of living relatives. The closest relatives of the hamadryas are the savanna baboons, *Papio anubis*, *P. cynocephalus* and *P. ursinus*. All investigated populations of savanna baboons, from Kenya to South Africa, are organized in multi-male groups with no internal fragmentation into one-male units. Since this one-level organization is also the general blueprint of the organizations found in the related genus of the macaques, we may reasonably assume that the organization of today's savanna baboons is the older type of baboon organization, from which the hamadryas pattern branched off as a more recent specialization. The present habitat of the hamadryas suggests that the specialization is somehow related to the survival in an arid environment.

In order to trace the course of this specialization, it is necessary to find at least some structural homologues in the organizations of savanna and desert baboons. The available evidence suggests that the hamadryas *band*, not the unstable troop, is the surviving homologue of the savanna baboon group. If this is true, two steps were necessary to create the additional levels – the one-male unit and the troop – which characterize the organization of modern hamadryas: 1. The original band or group became fragmented into one-male units, in that the females became permanently allocated to individual males of the band, probably through early adoption by the increasingly parental males. 2. The original bands or groups developed a mutual tolerance at least at the sleeping lairs. According to ALTMANN (personal communication), some groups of savanna baboons in the Amboseli area are close to realizing this second step: Groves of sleeping trees are scarce in their habitat, and several groups may assemble in the same grove at dusk, and separate again in the morning[1].

This course of evolution is, at first, somewhat surprising. From a superficial ecological examination, the transition from the savanna habitat to the semidesert appears as merely one more step in the direction which the ancestors of the savanna baboons initiated when they left the forests. In the hamadryas habitat, high retreats from predators are even scarcer than they already are in the savanna,

[1] Two important new publications should be considered in this context: ROWELL (1966) finds that the groups of her *forest* living anubis baboons are *open* and that they frequently exchange members. And CROOK (1966a) states that the one-male units of *Theropithecus gelada* form loose troops, but nothing resembling the integrated hamadryas band. The tendency to form closed multi-male units in open habitats thus may be an environmental modification typical of *Papio*, but not of *Theropithecus*. The hamadryas band may be a remnant of this tendency in the savanna baboons, whereas the gelada does not share this heritage, although otherwise his organization converges with that of the hamadryas. CROOK's paper includes a valuable comparison of gelada and hamadryas organizations.

and the diet is, at least in the dry season, even poorer and more dehydrated. The savanna baboons adapted to such conditions by evolving large, aggressive males who can jointly drive smaller predators away from the group, and by supplementing their diet by occasionally eating small mammals. The hamadryas, instead of proceeding further along this line of specialization, have taken a new course of evolution which suggests that the pressure from predators is less important in their habitat, whereas the nutritional situation becomes a powerful selective factor. This hypothesis is supported in three ways: 1. The relative size of the males in relation to females has not further increased in the hamadryas; the male-female weight ratio is about the same as it is in savanna baboons. 2. Both male and female hamadryas are considerably smaller than their savanna counterparts, suggesting that dietary problems outweigh the need for strong, big fighters. 3. The social reorganization of the hamadryas has partly abandoned the advantage which the closed group of the savanna type provided in the defense against predators. It has created small family groups with only one fighter, but it has thereby increased the species' capacity to disperse and to exploit the sparse and scattered food resources of its habitat.

Whereas sparseness of food seems to favor a population with small, dispersed social units, the scarceness of groves or cliffs for sleeping requires exactly the opposite, i.e., large concentrations of animals in a few spots. By changing from one to three levels of organization, the hamadryas apparently has specialized to alternately meet both needs. Its bands can split up into one-male units which may disperse during the day in search of food. On the other hand, bands can develop a relationship that permits them to share a sleeping rock, i.e., to form a troop. Thus, the populations have a capacity for organized fission and fusion which, according to our survey, is used in various combinations and degrees according to the local conditions[1].

While the hamadryas baboons clearly deviated from the line of specialization followed by their assumed savanna-living ancestors, they nevertheless retained several characteristics of the savanna baboons. The hamadryas still show the savanna baboons' sexual dimorphism as to size, but instead of increasing it, they added a dimorphism in color, which is most likely related to their social

[1] CROOK's (1966a) findings support this interpretation. Gelada one-male units form troops only in areas with good food supply; where food is scarce, they live apart from each other as independent units.

reorganization. Our ecological observations, which will be presented in the following chapters, further suggest that the hamadryas, like the savanna baboons, are occasionally carnivorous (p. 164), but again, this specialization is apparently not further increased in the hamadryas. Their survival in the semidesert areas may depend more on their habit of digging for water (p. 164), a behavior which has not been reported in the savanna baboons.

Socially, the change from one to several levels of organization creates two problems, both worthy of further research. First, some individuals must at the same time be members of at least two social units. Thus, in the hamadryas the adult male is both the leader of a unit and a member of the band, although at different stages of his life these roles receive different emphasis. Secondly, the multi-level organization involves a potential danger for the integrity of the lower units. This difficulty is, of course, not entirely new. Remaining together without mutual destruction is a problem which every society must solve on the level of its individuals. In the multi-level society, i.e., in a 'group of groups', the theme is merely repeated on the level of the component units. In both cases, spatial separation is no longer a permanent preventive against mutual aggression; at least in part, aggression must be controlled by inhibitions. Such inhibitions, however, do not entirely replace the techniques of spatial adjustments. In the hamadryas society, inhibitions are sufficient to permit proximity of units, but they cannot prevent aggression when unit spaces intersect. Spatial separation still functions as a last resort when the inhibitions break down, and as a routine among strangers, where no inhibitions exist.

VIII. INTERACTIONS WITH THE ENVIRONMENT

1. ECOLOGICAL OBSERVATIONS ON THE BROAD SAMPLE

The ecological conditions in the area of our broad sample (see Fig. 11) were surveyed by KURT. In the south, this area begins with the wooded foothills of the Ahmar mountains, standing about 100 meters above the northern plain. Towards the north, the foothills give way to bush country, which extends along the mountain foot in a band about 6 kilometers deep. Still, occasional hills rise about 20 meters from the plain. Because of its network of eroded furrows, this band was named the zone of the Wadis. Farther north, a more open and flat savanna follows, the *Acacia nubica*-band, which finally passes into the open grassland and semidesert of the Danakil plain.

The east-western extension of the area, from Erer-Gota to Garbelucu (also called Ocfalé) is about 20 kilometers. Within this distance, one permanent river (Erer) makes its way from a narrow ravine in the foothills through its flat bed in the plain to dry out in the semi-desert. Nine more riverbeds are flooded for only a few hours after heavy rains. At all other times, they provide smooth sandy roads used by baboons and men. The deep cracks in the earth of the Wadi zone, with their vertical banks, though rarely flooded, can only be crossed at certain points. Four mineral springs at the foot of the mountains form small ponds throughout the year.

Nine cliffs were used as permanent or temporary sleeping rocks within the region. All are nearly vertical and at least 15 meters high. Several lower and less steep cliffs remained unused. Without exception, the sleeping rocks border riverbeds, either in the form of 30-meter deep ravines in the lowest slopes of the Ahmar mountains, or, in the plain, as cliffs 15 to 25 meters high, rising above the outer bends of permanent or mostly dry rivers (Fig. 61). Caves were rare features in all cliffs. On the ravine rock, the cliff without caves was preferred to the opposite cliff which had two. The cave at the base of White Rock was never entered by the baboons. (For the only case where baboons withdrew into a cave, see page 113.) The average

Fig. 61. Central portion of the White Rock, the main location of the close study. Observations were made from the river bed in the foreground.

distance between neighboring sleeping rocks was 6.6 kilometers with 2 and 8 kilometers as the extremes.

The bridgeless all-weather road and the one-track railway from Addis Abeba to Diredawa cross the Wadi zone from west to east. About 10 cars and 4 trains passed each day.

A survey of the vegetation was made in the zones where hamadryas were usually encountered. Starting from the foothills, the zone of the Wadis was divided into zones I, II, and III, each of them extending for about 2 kilometers from south to north (Fig. 62). In each of these zones, 49 areas, at regular intervals from each other, were sampled. Each area was a 300 by 300 meter square. The most frequent types of trees and shrubs (as determined by Dr. W. Burger of the Alemaya Teachers Training College) were counted and their relative frequencies determined (Table XII). The composition of the undergrowth was judged by estimating the surface covered by each type (Table XIII).

In 1961, the short rainy season set in on March 11. For 6 weeks, one or two heavy downpours occurred almost daily. Within a few

Fig. 62. Typical vegetation in Wadi zone III photographed in the dry season, with the Ahmar mountains in the background. In the foreground are Sisal plants, whose leaf bases are eaten during this season.

Table XII. Percentages of the most frequent trees and shrubs in the 5 zones of the Erer-Gota habitat.

	Foot-hills	Wadi zones I	II	III	Gallery forest
Deciduous trees with short or no thorns	52	2	0.5	—	26
Shrub acacias (*A. nilotica Krausiana, A. laeta, A. sp.*)	25	21	41	20	37
A. nubica	—	—	11	24	1
Umbrella acacias	13	24	8	16	10
Large-leafed acacias	4	26	10	14	11
Dobera glabra (evergreen leaf tree)	5	27	30	26	14
Average distance between trees and shrubs	5 m	6 m	9 m	14 m	7 m

Table XIII. Estimated percentage of surface covered by different undergrowth plants in the 5 zones of the Erer-Gota habitat during the long rains.

	Foot-hills	Wadi zones I	Wadi zones II	Wadi zones III	Gallery forest
Grasses, typically *Digitaria*	7	14	38	51	29
Sisal	4	8	6	1	7
Euphorbia shrub	5	21	4	11	18
Herbs	58	14	7	—	32
Bare ground	26	43	45	37	14

days, the acacias of the southern zones sent out green leaves and then flowers, and later grass shot up knee-high and bloomed. The rains subsided for a few weeks in May while vegetation was at its height. On June 19, the long rainy season started with short heavy precipitations, flooding the dry river beds for hours to a degree where they often could not be passed by baboons. Swift currents up to 20 and 30 cm deep were still crossed by hamadryas parties in a series of jumps, the smaller animals being submerged up to their necks between jumps. Heavy rain would drive the baboons from the waiting areas into the shelter of the overhangs in the sleeping rock. When caught by downpours on route, they would sit quietly in the open, heads lowered and with the backs against the wind.

Early during the long rains the grass turned yellow. In September, the rains faded out with fine, day-long precipitations, and the dry season began. In January, the acacias had no leaves and the long grass was dry. In the plain, only the gallery forest remained green, while the Sisal, the Euphorbia shrubs and the typical *Dobera glabra* trees stood out as green patches in the savanna.

Hamadryas parties were encountered in the foothills, the three Wadi zones and, during the rains, in the *Acacia nubica* band. On several trips we did not see any in the arid northern grasslands.

On certain days, Kurt recorded the food eaten by each baboon he could observe at the moment he spotted the animal. Table XIV shows the frequencies of different kinds of food eaten by the baboons in the two main seasons. Though generally considered as extremely

Table XIV. Percentages of different food types eaten by the hamadryas of the Erer-Gota area.

Season	Food type	%	Way of gathering	Number of individuals sampled
End of dry season	Green leaves of deciduous trees other than acacias, in gallery forests	2	Climbing	485
	Beans and dry leaves of acacia	78	Climbing	
		6	Picking from ground	
	Sisal leaves	5	Tearing out of ground	
	Leaves of *Dobera glabra*	4	Climbing	
	Roots	4	Digging	
Long rains	Grass seeds	44	Picking from the ground	325
	Other food picked from the ground	8		
	Acacia flowers	43	Climbing	
	Fresh acacia shoots	3	Climbing	
	Fruits of *Dobera glabra*	3	Climbing	

terrestrial primates, our 'desert baboons' may be more arboreal in their feeding habits than the anubis baboons who sleep on trees but usually forage on the ground. Acacias, though deciduous, are seen to provide the main subsistence in both seasons. The soft, bitter base of the spear-like Sisal leaves and the thick waxy leaves of the evergreen *Dobera glabra* (Fig. 63), though available throughout the year, are only eaten in the dry season. Acacia flowers and grass seeds are preferred to any other type of food. Both are more abundant in the northern parts of the habitat. This is probably the reason why, during the rainy season, the daily routes pointed northwards from all rocks. Two rocks in the foothills were even entirely abandoned throughout the rainy season, and one troop from the mountains was seen in the northern plain at an unusually large distance from its home rock (p. 99). Strikingly, the permanently green, large (more than 15 meters) trees of the gallery forests were hardly used as sources of food and never as sleeping places. The baboons only approached them when they went to drink and rest at the waterholes in the rivers, and then only a few animals would occasionally climb onto limbs not higher than about 4 meters and pluck some fruits.

During the small rains, on the evening of April 5, swarms of locusts settled in our area. The next day, we observed a troop of hamadryas

Fig. 63. Baboons feeding from a *Dobera glabra* tree.

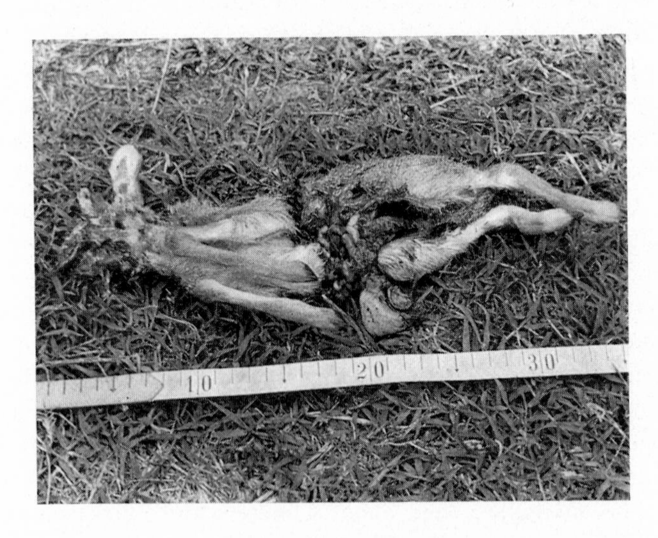

Fig. 64. A young Dik-dik antelope as it was abandoned by baboons. Measures are centimeters.

a

b

c

Fig. 65. (a) Two adult males drink from holes which they have dug in the sand. In the background, other baboons drink from a pool in the river. (b) Drinking holes dug by hamadryas baboons at the edge of a pool infested with algae. (c) Such holes, up to 20 centimeters deep, are often made far away from any open surface of water.

as they sat in the acacia shrub 80 meters from their sleeping rock, plucking the locusts from the twigs and eating large quantities of them. The baboons would either sort out everything but the wings and legs with tongue and lips, or they just bit off and ate the heads and abdomina. By 09.30 h most of the animals had stopped eating. They then set out in parties for a short meandering route which led them only 1.5 kilometers away from the rock.

Only one case of possible predation on small mammals was observed. On March 3, we spotted a 3-year-old hamadryas female hopping on three legs while carrying a dead young Dik-Dik antelope *(Madoqua kirki)* under her left arm, pressing its back alongside against her side. As she walked an adult male approached her and she fled into the bush, still carrying the Dik-Dik and screaming. We later found the Dik-Dik 20 meters away (Fig. 64). The abdominal cavity lay open, with organs complete; the facial skeleton was chewed up and partly missing, and on several spots on the side of the neck the skin was gnawed through. The carcass was fresh.

During the dry season, each troop had 2 to 4 permanent watering places within its range, mostly at pools under small chutes in the otherwise dry river beds. The hamadryas frequently dug individual drinking holes in the sand of the river beds (Fig. 65), at any distance from the natural pools, by pulling up the wet sand with their hands while sitting. In contrast to the water of the pools, the water collecting in such holes is cool and free of algae. In order to drink, the baboons rest their weight on their forearms, often anchoring their tails around a piece of rock. During the dry season, large parties of baboons regularly held prolonged noon siestas at the waterholes. In the wet months the animals would drink here and there as they crossed rivulets and streams.

Except for the last weeks of the long rains, there were several hours of sunshine every day. In the early morning, the hamadryas sat in the sun above the sleeping rocks, first facing the rays, later turning their backs. From 09.00 h onwards, they sat down in the shade of trees and bushes whenever possible, usually one one-male unit under a plant.

The mammal fauna of the Erer region included a few lions and cheetahs *(Acinonyx jubatus)*; spotted and striped hyenas *(Crocuta crocuta, Hyaena hyaena)*; jackals *(Thos adustus)*; earth wolves *(Proteles cristatus)*; wart hogs *(Phacochoerus aethiopicus)*; hyrax (species not determined); and among the antelopes the dik-dik

(Madoqua kirki), greater and lesser kudu *(Strepsiceros strepsiceros chora, S. imberbis)*, gerenuk *(Litocranius walleri)* and *Oreotragus*. Oryx *(O. beisa beisa)*, Soemmerings gazelles *(Gazella soemmering erlangi)* and ostriches *(Struthio camelus)* occur in the northern grassland. Only the large carnivores and the wart-hogs are hunted by man.

No contacts between predators and baboons were observed, and no evidence of such contacts was found except for a chorus of screams one night at the White Rock and the track of a large feline discovered next morning at the foot of the rock. Hamadryas were once observed 7 meters away from a greater kudu without any unusual behavior on either part. There were no signs of the symbiosis between ungulates and baboons in locating predators, observed by DeVore in Kenya, although the hamadryas once became alert at the nasal whistle of a Dik-Dik. Juvenile baboons sometimes lunged at the hyrax living on some sleeping rocks, but no physical contacts occurred. *Oreotragus* were sometimes seen on the northern rocks, together with the local troops. Only one interaction was seen between the two species: A 3-year old baboon approached an oreotragus, and sniffed its muzzle.

The area was inhabited by nine groups of the partly nomadic Gurgure people and their cattle. They were not seen to interfere with the daily routine of the hamadryas. Actually, the White Rock troop frequently walked 50 meters past one of the Gurgure settlements and fed on the grassy roof of some abandoned huts.

During the field year, Kurt (1964) studied the frequency and the migrations of the antelopes in our region. He found that most of them had more specialized food requirements than the baboons, and that only the small Dik-Diks, together with the baboons, held out in the Wadi zone throughout the dry season while the other species migrated according to the seasons. Reasons for these differences may be the baboon's ability to climb and forage on trees and, to a lesser degree, to dig for food and water. The thorny branches with which the nomads protect their own waterholes were sufficient to keep the wild ungulates away from the water, but the baboons were seen dragging the branches apart and drinking from the holes.

2. TIME-SPACE STRUCTURE OF THE DAILY ROUTES

The general characteristics of the daily travels through the feeding areas have been described on page 12. Along most of the route, a troop spreads out over nearly one square kilometer. Its actual outline is often lost from view, and the observer has to satisfy himself by following a party with which he manages to catch up. The routes shown in our maps thus represent an accidental line within the broad strip actually covered by the entire troop.

The numbered masts of the power line along the Addis Abeba-Diredawa road provided an axis for mapping. The distances between all masts were trigonometrically mapped with Recta field compasses. When walking with a troop, we determined our position approximately every 30 minutes.

Nine daily travels could be completely mapped in this way. One of them was the exceptionally short route (4.1 kilometers) of the two-male troop described on page 124. The length of the other eight

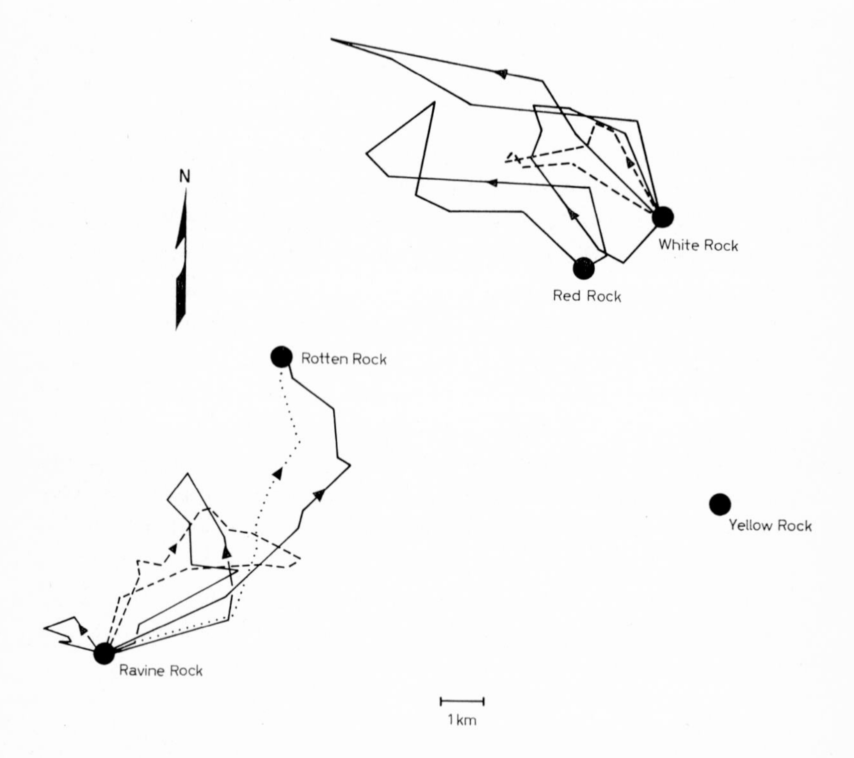

Fig. 66. Map of nine daily routes. The short route at the extreme left is that of Circum and Pater, described on page 124.

routes averaged 13.2 kilometers with a range from 9.8 to 19.2 kilometers, which is more than has so far been observed in other primates. Figure 66 shows the location of the travel routes in the map and their individual forms. In most of them, the return march parallels the outward march. In contrast, the shorter routes of anubis baboons in Nairobi usually follow a circular pattern (HALL and DEVORE, 1965) which is probably related to the higher density of sleeping places in their habitat. It is not accidental that most hamadryas routes from a given rock start out in the same direction. While each troop occasionally departed in any of a number of directions, one or two directions were chosen more frequently than all the others together. This may reflect the scarcity of the sleeping rocks as well as of dense groves of the main food plants, the acacias.

In the Erer area, the time of departure in the morning varied as much as 1 hour and 50 minutes in the same troop from one day to the next, and between 06.00 h and 10.00 h official time throughout the months from February through October. There is a positive rank order correlation of the morning schedule with the monthly average time of astronomical sunrise. The last animals left their sleeping ledge 28 to 38 minutes after sunrise ($R = 0.94$, $P < 0.05$), and the troop's departure from the waiting area followed the time of sunrise by 55 to 120 minutes in the monthly averages ($R = 0.83$, not significant). Not included in this calculation are the months of the long rains, July and August. During this time, on rainy as well as on sunny mornings, the animals waited twice as long before they left the cliff and before they departed from the area.

The duration of the daily march tends to be constant. When a troop departs before 08.00 h it will significantly more often be back at the rock before 17.00 h than when it departs later ($P = 0.03$). Thus, the schedule throughout the day is affected by the time of sunrise, while the time of sunset and darkness has no apparent effect on the time of return to the rock.

In Figure 67 our average speed when walking with the troops on the mapped routes was plotted against clock time and also against time relative to the moment of departure. The largest distance covered in a single hour was 4.7 kilometers. In the morning, the baboons sometimes crossed open areas in a slow gallop with an approximate speed of 8 kilometers per hour.

Since hamadryas troops are not constant in membership, their total range of travels in one year (the 'home range') cannot be

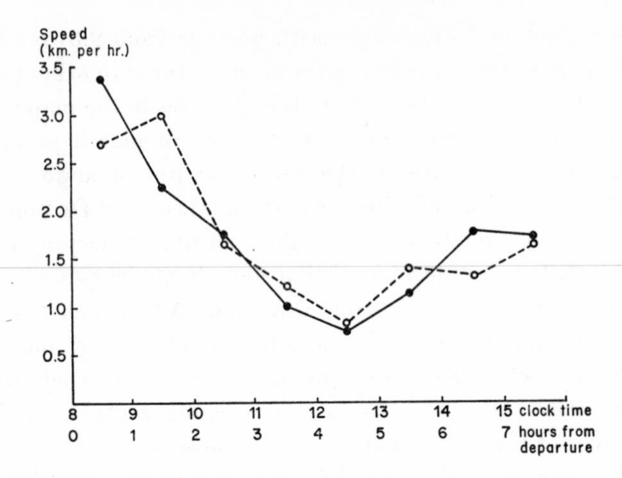

Fig. 67. Mean travelling speed per hour as measured in 8 daily routes. The broken curve represents the speed calculated for each hour of the day (clock time); for the solid curve, the hours were counted from the moment of departure.

determined. We know only that the routes of troops from different rocks overlap, and that one troop can occasionally penetrate right to the sleeping rock of its neighbors (p. 99 and 104). None of the fights between parties can be interpreted as a territorial dispute. The possibility of migrations was discussed on page 23.

3. RELATIONS BETWEEN LOCATION, ACTIVITY AND FLIGHT DISTANCE

Hediger (1951) introduced the term 'Fixpunkt' to designate the fact that certain animal activities are concentrated on certain spots in the home range. Two kinds of such action-specific locations were found in all regions visited: The sleeping rocks and, during the dry season, the water holes. Table XV shows the distribution of behavioral categories over the different locations. For most categories a gradient of occurrence is observed from left to right in the table, the two sets of behavior observed at the sleeping rocks and on route being most different from each other and more specific than the intermediate set recorded at the water holes. In the vertical dimension of the table, the behaviors most specific to a location are found on top and at the end: The baboons did not eat food (other than the corn we gave them

Table XV. Occurrence of main behavioral categories at different locations in the home range. + = regularly observed, (+) = rare, — = no records.

	At sleeping rocks and waiting areas	At water holes	On route
Eating	—	(+)	+
Drinking	—	+	+
Grooming	+	+	+
Neck biting	+	+	+
Playing	+	+	(+)
Fighting	+	—	—
Copulation	+	—	—
Sleeping	+	—	—

on a few occasions) at or around their sleeping rock although it was available there, and they did not sleep, copulate or fight on route.

Social behavior in general was concentrated at the 'Fixpunkte' for sleeping and drinking, preceding and following the location-specific action. In the dry season, it was striking to see that the large parties gradually assembling at a water hole did not drink at once but first settled down for grooming and play. On 6 occasions when the times were measured, an average of 69 minutes elapsed before the first baboons drank, the drinking being followed each time by another social period.

A general motivational change probably occurs when a party gathers to rest at a spot of vital importance. The change in unit formation described on page 123 is an indication of this, as is the significant change of the distance at which the observer is tolerated at various places. We can tentatively describe a 'resting mood' which facilitates grooming, sexual behavior, fighting and play, and a 'travelling mood' which facilitates foraging and flight while inhibiting most types of social interactions.

The distances at which the troops tolerated the observers changed predictably in the course of the daily routine. The values given in Table XVI were collected during the first days spent with each of 5 troops in the Erer-Gota region, i.e. before they had become fully habituated to our presence. Later, we avoided making the test, trying to stay beyond these distances. Distances were estimated at the moment when the seated party closest to the visibly and slowly approaching observer stopped its activities, got up and walked away.

Table XVI. Flight distances (meters) towards observers at different locations in the home range. n = number of estimates.

At sleeping rock evening	At sleeping rock morning	First 300 m after departure	On route morning	At water hole before drinking	At water hole after drinking	On route at turning point	On route in the afternoon	Waiting area evening
30±1.6	51±7.0	158±34.4	63±5.4	40±2.2	52±3.7	59±5.0	55±4.5	41±3.4
n = 10	n = 5	n = 7	n = 32	n = 14	n = 10	n = 24	n = 10	n = 11

The estimates were frequently checked, both by pacing off the distance and with the tape measure. Agreements between observers were within 5 meters at distances less than 60 meters. T-tests were applied to compare flight distances in different situations, the level of significance being set at $P = 0.001$.

As Table XVI shows, the flight distance was lowest when the troop had settled down on the rock in the evening. In the morning, before departure, it became significantly higher and rose sharply when the troop had departed from the rock. On route it seemed to decrease throughout the day, but the difference between morning and afternoon is not significant. At the waterhole, around noon, there was a temporary and significant reduction in flight distance, and, as at the sleeping rock, the distance was significantly lower before the location-specific action than after its occurrence. Habituation to our presence occurred within the first two weeks and reduced the figures by approximately one-half, but the daily cycle of changes was still obvious. Flight distances were and remained significantly higher in two troops near Erer-Gota, where the baboons were often chased and molested by villagers. Even there, however, location had its effect: The flight distance at the rocks in the morning was 113 ± 37 meters and rose to 383 ± 37 meters after departure.

Apparently, the flight distance is not only a response of the animals to the observer, but also a response to the location. If the first variable is kept constant, flight distance becomes a measure of the animals' relationship to the various parts of their home range. The data encourage an intensive study, which would preferably be done after the habituation period. In addition to measuring the distances kept at each of a large number of locations in a home range, the observer should also try to approach the parties from different directions at each location, especially near the 'Fixpunkte' and near the border of the range. Such a study might result in a vectorial map of the degree and direction of travelling tendencies and in an understanding of the 'field characteristics' of the home range. The type of behavior occurring at the moment and the time of day would have to be considered as further independent variables.

4. MANIPULATION OF OBJECTS

The hamadryas baboons rarely handled any nonedible objects. The highest frequency was reached by 1-year-olds who did so in 3.5% of observation minutes at the rock (KUMMER and KURT, 1965). The type of objects most handled were stones from 4 to 10 cm in diameter. Young juveniles sometimes turned such stones during the rest periods on the sleeping rock and then looked into the appearing hollow, but we never saw them pick up and eat anything. Such activity usually attracted other juveniles who then started turning stones close to the first animal.

Juveniles were seen standing on the upper edge of the sleeping cliff, loosening stones by pressing them out with one hand and watching as they fell down the rock-face. Here again, their peers tended to join in. In the same way, an adult male leisurely loosened and watched stones falling as he was standing near the edge while being groomed. The second recorded manipulation by an adult also occurred while the animal, a female, was being groomed in standing position. She grasped a forked dry twig, held it for a few seconds in her outstretched hand and gazed at it, then flung it 50 cm away with a medial swing of the arm. Twigs and stalks of grass were sometimes chewed and carried in the mouth by black infants.

Stones were frequently loosened accidentally by baboons climbing in the face of the rock and were carefully avoided by others. One morning[1], an adult male stepped towards the edge of the White Rock. As he sat down, a stone of about 6 cm broke loose from under him. On a ledge directly underneath, about 2 meters away, three one-year-olds were wrestling. The adult male grabbed the rolling stone and held it, with his hand resting palm down on the knee. A female came

[1] This sequence was observed by the film producer, Mr. AUGUST KERN.

over and groomed him while he watched the infants. After nearly one minute, the infants moved on and the male dropped the stone which fell down the cliff across the abandoned playing ground. The baboons never dropped or threw any objects at us.

The longest sequence of manipulation was demonstrated by a two-year-old female who carried, hugged and protected a stone for at least 12 minutes.

18.11 h The male Fork is leading his unit down the cliff towards the sleeping ledge. His 2-year-old female is seen carrying a white round stone about 5 cm in diameter, pressing it against her chest with one hand as she climbs down. She thus struggles over a difficult passage. On the ledge, a one-year-old male chases her and she escapes, hugging the stone. During a short play-biting sequence with two juvenile males she is lying on her back, still holding the stone against her chest. For a while the males withdraw and the female sits up, hugs the stone with both arms and touches it lightly with her mouth. As a one-year old male comes back to her, she holds the stone closer, and threatens him by raising her brows, then escapes with the stone pressed against her chest.

18.20 h She then lies down on her belly and holds the stone with both hands in front of her face, watching it. She sits up and rolls it with both hands against the ground, then gently strokes its surface, with both hands moving sideways over it. During a short play-biting sequence with the one-year-old male she abandons the stone, then picks it up again and carries it on her chest farther down the cliff.

18.23 h The stone slips out of her embrace and falls down the cliff while she and the one-year-old male watch its fall.

No artificial objects thrown away by humans were found in the area used by the White Rock troop. In an attempt to mark juveniles at the sleeping cliff, we distributed plastic bottles 4 cm in length and filled with dye before the troop arrived. After sniffing them, the juveniles manipulated and carried the bottles, smearing themselves with the dye. Kernels of corn, though unknown to the baboons, were readily eaten when scattered by us, but half a dozen peeled corn-cobs distributed over the White Rock were sufficient to keep the arriving troop away from the cliff for 32 minutes.

The parties sit waiting in the river bed, 30 meters away from the cobs, giving the alarm bark. Finally, a female presents to her male. He responds by mounting and frantically thrusting, and then rushes with her to the foot of the cliff. Another leader at once mounts one of his females and advances 3 meters to the rock. Presently, many females and juveniles press forward for a few meters but stop when the adult males do not follow them.

Playing juveniles are the first to enter the rock area, and two minutes later the first one-male units follow them. As the troop gradually moves in, most animals stop to sniff the cobs. Later, adult males and juveniles pick up cobs, tear at them with their teeth, but do not eat them. Juveniles also carry the cobs while chasing each other.

5. CONTACTS BETWEEN HAMADRYAS AND GELADAS

As mentioned briefly on page 23, parties of hamadryas baboons were seen mixing with parties of *Theropithecus gelada* in a valley named Aliltu (39° 10′ E. Long.; 9° 9′ N. Lat.), considerably more to the south than the southernmost locality of the gelada range (Magdala) mentioned by Tappen (1960). Contacts were observed throughout the day of December 22. After passing the night on the same cliff, but about 100 meters away from the geladas, the hamadryas travelled down the valley and disappeared kilometers away in the direction of the Danakil plain to the east. During a visit in January, no hamadryas were living in the valley.

The hamadryas troop numbered about 55 animals, 6 of which were fully adult males. At times they formed two or three detached parties, each feeding on one of the threshing places distributed over the steep, cultivated slopes of the valley. At most times, each hamadryas party was accompanied by a party of the gelada troop which numbered more than 80 animals and included 8 adult males. Such mixed feeding was observed four times during the day. The hamadryas parties each time arrived first and settled to eat the grain, while the accompanying geladas idly waited 20 to 100 meters away without eating. In one case, the geladas only moved in after the hamadryas had left. On the other three occasions, they joined the hamadryas on the threshing grounds which measured about 15 meters in diameter. One instance was observed in detail:

A hamadryas party is feeding on the ground with one adult male in the center of the ground and 4 others on the party's periphery, outside the actual threshing floor. 11 geladas are sitting still farther away without access to the food. From the opposite direction, another gelada party of 33, with two adult males, hesitatingly approaches the place. The adult gelada male in front sits down facing the closest hamadryas male at 8 meters. The rest of the arriving geladas sit down behind him. Seconds later, the second of the arriving gelada males rises, passes the first as well as the hamadryas male, then runs to the edge of the feeding ground where he starts to eat. As the gelada females and juveniles close in, the hamadryas male, who did not respond so far, lunges at them. Most of them flee, but two females rush past the hamadryas male and join the gelada male in the center. Within five minutes, the geladas infiltrate and the parties mix. An excited male-female mounting and a neck-bite on one of the hamadryas females are the only responses on the side of the hama-dryas.

During the entire day, no other aggressive acts between the two genera were observed. They intermingled at the feeding places, one adult gelada male sitting and eating only one meter from an adult hamadryas male. The hamadryas males generally occupied the central parts of the feeding ground, while the gelada males more often sat on the periphery and outside the feeding place. The average of the estimated distances from the center was 5.7 meters ($n = 17$) for the hamadryas males and 12.4 meters ($n = 14$) for the gelada males.

Twice, the geladas and the hamadryas travelled together in a closed column for more than one kilometer. Here, the genera were not riixing, but the hamadryas took the front and led the way, while the mladas followed closely. Both parties usually had an adult male in fgont and in the rear. When the hamadryas headed toward the erver to drink and rest, the geladas stayed 150 meters behind and waited rather than drinking at another place. When the hamadryas finally left the river after a siesta of one and a half hours, the geladas immediately came down to drink in the same place.

On three occasions, mixed feeding parties fled from approaching Amhara people. In one case, the geladas fled first and were followed by the hamadryas who, however, moved up to the front when the escape slowed down. On the other two occasions, the two genera separated and fled in opposite directions. The geladas travelled alone for some time, but later one of their parties rejoined and followed the hamadryas troop on the long way to the drinking place.

In general, the geladas seemed to join the hamadryas more often than vice versa. Taken together, the observations suggest that the hamadryas were 'dominant' over the geladas. This would appear to be a prerequisite for their intrusion on the gelada habitat during the dry season (p. 23), and it supports TAPPEN's (1960) view that geladas as a species seem to occupy a refuge habitat whose altitude enables them to survive the competition with baboons.

IX. REPRODUCTION

1. BEHAVIOR AT BIRTH

The only birth actually observed occurred at dusk, shortly after 18.00 h on September 7, on a sleeping ledge of the White Rock. At the moment of birth the mother was squatting (a posture which we never saw in other contexts) 2 meters away from the rest of the unit, facing away from them. She then turned away from us, and we did not see whether she cleaned the infant. Some 10 minutes later, she moved a few steps on three legs carrying the placenta in her mouth and supporting the infant at her belly with one hand, the umbilical cord passing between her arms. She then sat down again and began to chew on the edge of the placenta for 20 minutes until darkness. Occasionally, she held the placenta with one or both hands while chewing it, but she did not tear it and its size did not visibly diminish. Our impression was that the placenta was not eaten, but sucked. The infant once released its grip on the mother's flanks, and for a short time was lying on the ground fidgeting with arms and legs. Several adult females passed by and took a short look. A two-year-old male sat down close to the mother shortly after birth. Later, another male of the same age joined him and both males intently watched the mother until dark and followed her when she moved. The unit leader never approached the mother.

During the night of August 10 a female of the mothers' unit had given birth. As she left the sleeping cliff in the morning with her newborn infant clinging to her belly, the umbilical cord with the dry placenta was still attached to the infant. The placenta was dragged along between the mother's feet and visibly interfered with her walking.

2. SYNCHRONIZATION OF REPRODUCTIVE CYCLES

The following data were collected on 7 troops of the broad sample from January to October, 1961. The reproductive states of females in walking troops and parties were recorded at irregular intervals.

Three classes of females were distinguished: Females carrying black infants, females with perineal swellings larger than about 5 cm in diameter and females having neither characteristic, i.e., anoestrous, pregnant or lactating females. In another type of count, the reproductive states of detached one-male units were recorded. The per minute frequencies of sexual mountings with intromission were recorded in a total of 99 sessions lasting from 40 to 120 minutes at the sleeping rocks, and the resulting figures divided by the number of the baboons which were in full view.

Data on the duration of the reproductive phases are only available for captive hamadryas. ZUCKERMAN and PARKES (1930) found an average oestrous cycle of 30 days with 23 and 40 days as the extremes. The perineal swelling develops in 5 to 7 days and remains at its height for about one week. The average gestation period was determined by ZUCKERMAN (1953) as being 170 days for the London colony, and by LAKIN (1959) as 173 days for the hamadryas females at the Sukhumi Station in Russia. The period of lactation varies widely. For 15 females an average of 239 days is reported by BOCHKAREV (1933). Five to eight days after the end of lactation the menstrual cycles started again. HALL and DEVORE (1965) estimate that the infants of free-living savanna baboons are weaned only at the age of 11 to 15 months.

In a study on the seasonal distribution of 92 births in the Zoos of London, Munich and Giza (Egypt), ZUCKERMAN (1953) found that most hamadryas infants were born between October and December. He concluded that a period of increased fertility occurs between May and July. In spite of the difference in climate, the population of Erer-Gota actually showed a pronounced peak of mating activity at this time (Fig. 68), as measured in the frequencies of swellings and sexual mountings. In addition, the swellings were at least double the size of those observed between January and April.

At the same time, in May and June, the Erer-Gota population also showed a marked birth peak, as measured by the frequency of females with black infants. This leads to the conclusion that a second joint peak of mating and births occurs in November and December. During our travels in December we had actually recorded a 'high number of black infants'. Most births would thus occur around the onset of the long rains and again in the middle of the dry season.

Repeated counts in June showed that the 6 troops included differed in the time of their birth peak. For instance, no less than

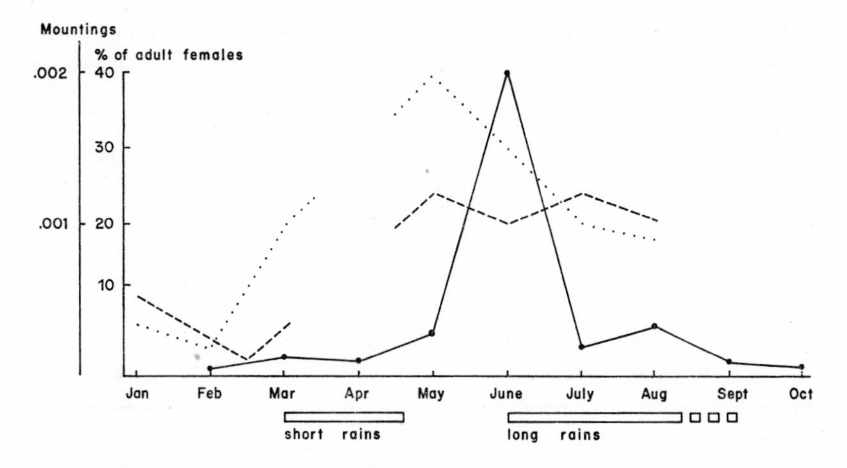

Fig. 68. Reproductive states in the broad sample, January through October 1961. Solid: Rate of sexual mountings per animal and minute. Broken: Percentage of adult females with black infants. Dotted: Percentage of adult females with swellings.

45% of the adult females at Red Rock had black infants at the end of May, while a less pronounced birth peak (24%) at White Rock, only 1.8 kilometers away, occurred two months later. (This produced the second peak in the curve of Fig. 68.)

A broad count in June was started when we had noticed that the females of a one-male unit very often showed the same phase of sexual swelling. For Table XVII only the detached units containing 2 adult females were selected in order to compare the observed combination of reproductive states in units with the calculated expectation under conditions of random combination. The expected values were calculated from the frequencies of the three types of females in the sample. It is obvious that females of different reproductive states were not combined at random. If one female of a unit was in oestrus, the probability was high that the other female was found in the same state. The majority of one-male units belonged to two types: Those with all females in oestrus and those where all females either had a black infant or were otherwise sexually inactive. Only 5% of the units instead of the expected 42% combined both conditions. Larger units, of course, showed more frequent exceptions to the rule. If they are included, as much as 17% of the units combined both conditions. Two- and three-year old females followed the same pattern as the adults. In troops with many adult females in oestrus, the percentage of juveniles in oestrus was also high. The

Table XVII. Combinations of reproductive states in detached one-male units with 2 adult females in June, 1961. Observed frequencies compared with frequencies expected in case of random combination. S = female with swelling; M = mother with black infant; O = female with none of both.

	Number of units							Number of females			
	OO	SS	MM	OS	OM	SM	Total	O	S	M	Total
Police rock	4	0	0	2	3	0	9	13	2	3	18
Cone rock	1	3	0	0	1	0	5	3	6	1	10
Rotten rock	4	1	0	0	0	0	5	8	2	0	10
Red rock	1	1	2	0	5	0	9	7	2	9	18
White rock	6	6	0	0	1	0	13	13	12	1	26
Total frequencies:											
Observed	16	11	2	2	10	0	41	44	24	14	82
Expected	12	4	1	13	7	4	41				
Total percentages:											
Observed	39%	27%	5%	5%	24%	0%	100%	53.7%	29.3%	17.1%	100.1%
Expected	29%	9%	3%	32%	18%	10%	101%				

two- and three-year-olds were as often in accordance with the rest of their one-male unit (75% and 69%) as were the adult females (78%).

After the birth and mating peak, many units lost their oestral uniformity. Whereas in May and June, 83% of 90 units with more than one female were uniform, the figure dropped to 61% in the 54 units counted from July to September. The decrease was primarily due to the increasing number of females without swelling in units of oestrous females. These units either must have lost the synchronicity of their cycles or, more probably, an increasing number of their females became pregnant.

Clearly, our data show merely a few aspects of the complex situation: First a birth peak and a mating peak were observed at the same time, suggesting that two different seasonal cycles occur in the same population. It is not known whether a given female will shift from one birth peak to the other in successive years, or whether each one-male unit or band consistently adheres to one birth peak. Secondly, some troops can have many newborn infants but few females with swellings while neighboring troops show an opposite ratio. The birth peaks of two neighboring troops occurred two months apart although their sleeping rocks faced the same direction and their overlapping home ranges were of similar quality. Both local diseases and social factors might be responsible.

The importance of social factors is evident in the synchronization of the female cycles in one-male units. Not only is a *monthly* cycle synchronized here, but also the females of a unit are usually found in the same *seasonal* cycle. Of the many possible interpretations none can be supported by these preliminary data. Seasonal cycles may be synchronized by adjusting the duration of the lactation period. In the monthly cycle, a 'unit oestrus' may be caused by the behavior of the leader or by mutual stimulation of the females within, but not across the units. Whatever the causes, the observations of synchronized troops and one-male units demonstrate that within the range set by the environment and the genotypes of our small baboon population, variation did not occur so much between individuals as between entire social entities who followed their private schedule regardless of those of their neighbors.

X. LIST OF PATTERNS OF SOCIAL BEHAVIOR

This is an informal catalogue of the behavior patterns used in the descriptions of this study, neglecting the variations. Patterns are crudely interpreted and grouped according to their tendency to occur together. The order within each group is that in which patterns are observed in complete sequences of rising intensity. If technically possible, patterns of lower intensity often persist throughout the sequence.

One asterisk marks all the patterns counted as 'social interactions' in our quantitative analyses. Two asterisks mark prolonged contacts of which only the beginning was counted as a 'social interaction'.

General

Looking at partner
Sniffing mouth (Fig. 47)
Approaching or following
Walking away or avoiding
*Pulling closer
*Shoving away

Aggression

Staring (Fig. 17 a)
*Raising brows (Fig. 41 d)
*Protruding head (Fig. 44)
*Slapping ground
*Pumping cheeks with chewing movements
*Opening mouth (at highest intensity, teeth are bared, Fig. 20 b)
*Lunging at partner (Fig. 20 a)
*Biting on shoulder (Fig. 19)
'Oohu' roar (by onlookers only)

Submission, escape

*Presenting rear
*Baring teeth with closed jaws
Staccato-coughing

Screaming, squealing
*Bending elbows and knees while standing
*Crouching (pressing belly to the ground)
*Escaping

Herding
*Aggressive patterns, excluding shoulder-biting
*Biting female's nape of the neck or back (Fig. 18)
*Crouching over female (complete continuum between this and mounting, Fig. 41 b)

Behavior between aggressive acts ('tension')
Undirected 'yawning'
*Mounting (Fig. 41 c)

Maternal behavior
*Inspecting (putting nostrils close to)
*Grabbing (Fig. 48 a)
**Embracing (Fig. 24)
*Presenting lowered hindquarters to infant as an invitation to carry him (Fig. 31 a)
**Carrying on belly or back (Fig. 23 and 45 b)

Infantile behavior
**Clinging, being carried
**Sucking
Humming (when left by mother)

Sexual
*Presenting anogenital region (female, Fig. 17 b)
*Inspecting and touching anogenital region (male)
*Putting hands on partner's flank (male)
*Mounting and copulating
*Being mounted

Comfort, appeasement
Contact grunt
*Tongue and lip smacking
*Presenting part of body to be groomed
*Being groomed

*Grooming (Fig. 32)
**Cuddling together (Fig. 16)

Play

*Wrestling (Fig. 26)
*Biting (inhibited)
*Chasing
*Being chased

Notifying among adult males

*Presenting face (Fig. 52 a)
*Presenting anal field (Fig. 52 b)
*Touching genitals

Uncertainty, crisis

Scratching
Touching hand to muzzle
Undirected 'yawning' (Fig. 22)
'Bahu' bark

XI. REFERENCES

ALTMANN, S.: A field study of the sociobiology of rhesus monkeys, *Macaca mulatta*. Ann. N.Y. Acad. Sci. *102:* 338–435 (1962).

BOCHKAREV, P. V.: Materials on the study of female sexual system physiology in monkeys. Arkhiv biologicheskikh nauk *33* (1–2): 263–269 (1933). Cit. in: The baboon, an annotated bibliography, p. 40 (Southwest Found. Research and Education, San Antonio, Texas 1964).

BOWDEN, D.: Primate behavioral research in the USSR – The Sukhumi Medico-Biological Station. Folia primat. *4:* 346–360 (1966).

CARPENTER, C. R.: Social behavior of nonhuman primates. In: CARPENTER, C. R.: Naturalistic behavior of nonhuman primates, pp. 365–385 (Pennsylvania State Univ. Press 1964).

CHANCE, M. R. A.: An interpretation of some agonistic postures; the role of 'cut-off' acts and postures. Symp. Zool. Soc. Lond. *8:* 71–89 (1962).

CHANCE, M. R. A.: Resolution of social conflict in animals and man. In: Ciba Found. Symp. on Conflict in Society. Ed. A. V. S. DE REUCK (Churchill Ltd., London 1966).

CROOK, J. H.: Gelada baboon herd structure and movement – A comparative report. Symp. Zool. Soc. Lond. 18: 237–248 (1966a).

CROOK, J. H. and GARTLAN, J. S.: Evolution of primate societies. Nature *210:* 1200–1203 (1966).

DEVORE, I.: The social behavior and organization of baboon troops. Unpublished doctoral dissertation. Dept. Anthropology, U. of Chicago (1962).

ELLIOT, D. G.: A review of the primates. Amer. Mus. Nat. History, Vol. 2 (1913).

HALL, K. R. L.: Numerical data, maintenance activities and locomotion of the wild chacma baboon, *Papio ursinus*. Proc. Zool. Soc. Lond. *139:* 181–220 (1962a).

HALL, K. R. L.: The sexual, agonistic and derived social behavior patterns of the wild chacma baboon, *Papio ursinus*. Proc. Zool. Soc. Lond. *139:* 283–327 (1962b).

HALL, K. R. L. and DE VORE, I.: Baboon ecology. Baboon Social Behavior. In: Primate Behavior, pp. 20–110. Ed. I. DEVORE (Holt, Rinehart and Winston, New York 1965).

HALL, K. R. L.: Behaviour and ecology of the wild Patas monkey, *Erythrocebus patas*, in Uganda. J. Zool. *148:* 15–87 (1965).

HARLOW, H. F.: The heterosexual affectional system in monkeys. Amer. Psychologist *17:* 1–9 (1962).

HEDIGER, H.: Zur Biologie und Psychologie der Flucht bei Tieren. Biol. Zentralbl. *54:* 21–40 (1934).

HEDIGER, H.: Observations sur la psychologie animale dans les parcs nationaux du Congo Belge (Instituts des Parcs Nationaux du Congo Belge, Bruxelles 1951).

HEINROTH, K.: Beobachtungen an handaufgezogenen Mantelpavianen (*Papio hamadryas* L.). Z. Tierpsychol. *16:* 705–732 (1959).

Hurme, V. O. and van Wagenen, G.: (Data on tooth eruption in *Macaca mulatta.*) In: Growth, pp. 403–404. Ph. L. Altman and D. S. Dittmer, Editors. Federation of American Societies for Experimental Biology (1962).

Iersel, T. T. A. van and Bol, A.: Preening of two tern species. A study of displacement activities. Behaviour *13:* 1–88 (1958).

Itani, J.: Paternal care in the wild Japanese monkey, *Macaca fuscata.* In: Primate Social Behavior, pp. 91–97. Ed. C. H. Southwick (van Nostrand, Princeton 1963).

Kummer, H.: Soziales Verhalten einer Mantelpavian-Gruppe. Beiheft 33 Schweiz. Z. Psychol. (Huber, Bern 1957).

Kummer, H. and Kurt, F.: Social units of a free-living population of hamadryas baboons. Folia primat. *1:* 4–19 (1963).

Kummer, H. and Kurt, F.: A comparison of social behavior in captive and wild hamadryas baboons. In: The baboon in medical research, pp. 65–80. H. Vagtborg, Editor (Univ. Texas Press 1965).

Kummer, H.: Tripartite relations in hamadryas baboons. In: Social interactions among primates. S. Altmann, Editor (Univ. Chicago Press 1967).

Kummer, H.: Two variations in the social organization of baboons. In: Primates: Studies in adaptation and variability. Ph. Jay, Editor (Holt, Rinehart and Winston, in press).

Kurt, F.: Beobachtungen an ostaethiopischen Antilopen. Vierteljahrschr. Naturforsch. Ges. Zürich *109:* 143–162 (1964).

Lakin, G. F.: Duration of pregnancy in hamadryas baboons and some other lower monkeys. Sovetskaja antropologiya *3* (1): 49–55 (1959). Cit. in: The baboon, an annotated bibliography (Southwest Found. Research and Education, San Antonio, Texas 1964).

Lorenz, K.: Das sogenannte Böse (Borotha-Schoeler, Wien 1963).

Lorenz, K.: King Solomon's Ring (Thomas Y. Crowell Co., New York 1952).

MacDonald, J.: Almost human (Chilton, Philadelphia 1965).

Reed, O. M.: Studies of the dentition and eruption patterns in the San Antonio Baboon Colony. In: The baboon in medical research, pp. 167–180. H. Vagtborg, Editor (Univ. Texas Press 1965).

Rowell. T. E.: Forest living baboons in Uganda. J. Zool. Lond. *149*: 344–364 (1966).

Starck, D. und Frick, H.: Beobachtungen an aethiopischen Primaten. Zool· Jahrb., Abt. Syst., Oekol. u. Geogr. d. Tiere *86:* 41–70 (1958).

Sugiyama, Y.: An artificial social change in a hanuman langur troop (*Presbytis entellus*). Primates 7: 41–72 (1966).

Tappen, N. C.: Problems of distribution and adaptation of the African monkeys. Current Anthropology *1:* 91–120 (1960).

Waerden, B. L. van der: Mathematische Statistik (Springer, Berlin 1957).

Washburn, S. L. and DeVore, I.: The social life of baboons. Sci. Amer. *204:* 62–71 (1961 a).

Washburn, S. L. and DeVore, I.: Social behavior of baboons and early man. Viking Fund Publ. Anthrop. *31:* 91–105 (1961 b).

Washburn, S. L. and DeVore, I.: Ecologie et comportement des babouins. La Terre et la Vie *2:* 133–149 (1962).

Young, W. C.; Goy, R. W. and Phoenix, C. H.: Hormones and sexual behavior. Science *143:* 212–218 (1964).

Zuckerman, S. and Parkes, A. S.: The oestrous cycle of the hamadryas baboon. J. Physiol. (Lond.) *69:* XXXI (1930).

Zuckerman, S.: The social life of monkeys and apes (Kegan Paul, London 1932).

Zuckerman, S.: Breeding seasons of mammals in captivity. Proc. Zool. Soc. Lond. *122:* 827–950 (1953).

XII. Subject Index